W9-BGP-307

The Unreasonable Effectiveness
of Number Theory

Recent Titles in This Series

(Continued in the back of this publication)

AMS SHORT COURSE LECTURE NOTES
Introductory Survey Lectures
published as a subseries of
Proceedings of Symposia in Applied Mathematics

Proceedings of Symposia in
APPLIED MATHEMATICS

Volume 46

The Unreasonable Effectiveness
of Number Theory

Stefan A. Burr, Editor

George E. Andrews
J. C. Lagarias
George Marsaglia
M. Douglas McIlroy
Vera Pless
Manfred R. Schroeder

American Mathematical Society
Providence, Rhode Island

LECTURE NOTES PREPARED FOR THE
AMERICAN MATHEMATICAL SOCIETY SHORT COURSE

THE UNREASONABLE EFFECTIVENESS OF NUMBER THEORY

HELD IN ORONO, MAINE
AUGUST 6–7, 1991

The AMS Short Course Series is sponsored by the Society's Program Committee for National Meetings. The series is under the direction of the Short Course Subcommittee of the Program Committee for National Meetings.

Library of Congress Cataloging-in-Publication Data

The unreasonable effectiveness of number theory / Stefan A. Burr, editor; George E. Andrews ... [et al.].
    p. cm. — (Proceedings of symposia in applied mathematics, ISSN 0160-7634; v. 46. AMS short course lecture notes)
    Course held in Orono, Maine, Aug. 6–7, 1991.
    Includes index.
    ISBN 0-8218-5501-8 (hard bound; acid free)
    1. Number theory–Congresses. I. Burr, Stefan A. (Stefan Andrus), 1940–. II. Andrews, George E., 1938–. III. Series: Proceedings of symposia in applied mathematics; v. 46. IV. Series: Proceedings of symposia in applied mathematics. AMS short course lecture notes.
QA241.U67 1992   92-24328
$512'.7$–dc20   CIP

COPYING AND REPRINTING. Individual readers of this publication, and nonprofit libraries acting for them, are permitted to make fair use of the material, such as to copy an article for use in teaching or research. Permission is granted to quote brief passages from this publication in reviews, provided the customary acknowledgment of the source is given.

    Republication, systematic copying, or multiple reproduction of any material in this publication (including abstracts) is permitted only under license from the American Mathematical Society. Requests for such permission should be addressed to the Manager of Editorial Services, American Mathematical Society, P.O. Box 6248, Providence, Rhode Island 02940-6248.

    The appearance of the code on the first page of an article in this book indicates the copyright owner's consent for copying beyond that permitted by Sections 107 or 108 of the U.S. Copyright Law, provided that the fee of $1.00 plus $.25 per page for each copy be paid directly to the Copyright Clearance Center, Inc., 27 Congress Street, Salem, Massachusetts 01970. This consent does not extend to other kinds of copying, such as copying for general distribution, for advertising or promotional purposes, for creating new collective works, or for resale.

1991 *Mathematics Subject Classification.*
Primary 11-06, 11K45, 11T71, 11Z50.
Copyright © 1992 by the American Mathematical Society. All rights reserved.
Printed in the United States of America.
The paper used in this book is acid-free and falls within the guidelines
established to ensure permanence and durability. ∞
Portions of this volume were printed
directly from author-prepared copy.
Portions of this volume were typeset by the authors
using $\mathcal{A}\mathcal{M}\mathcal{S}$-TEX, the American Mathematical Society's TEX macro system.

10 9 8 7 6 5 4 3 2 1   97 96 95 94 93 92

512,7
Un7

Table of Contents

Preface

In August 1991, the American Mathematical Society sponsored a Short Course at the summer meeting in Orono, Maine, entitled "The Unreasonable Effectiveness of Number Theory". Two years earlier, another Short Course was held on "Cryptology and Computational Number Theory", which emphasized cryptologic applications. Therefore, the Short Course in Orono concentrated on the great breadth of applications outside cryptology. This volume is based on the lectures given at that Short Course.

Number theory is one of the oldest and noblest branches of mathematics; indeed, it was already ancient in the time of Euclid. In fact, for almost all of its history it has seemed to be among the purest branches of mathematics. It is only within the last few decades that a large number of applications have been encountered, at least by the mathematical community. The applications to cryptology are now famous; but it is not as well known that number theory has found an enormous number and variety of real-world applications in many different fields. Indeed, the standard Mathematics Subject Classification includes several codes devoted entirely or heavily to applications of number theory.

What are the sources of these applications? The largest impetus has been, not surprisingly, the computer, with its digital and numerical nature. As with cryptology, the computer has been a driving force in the development of algebraic coding theory, random number generation, raster graphics, computer arithmetic, fast transforms, and many other areas. Perhaps surprisingly, physics, in spite of its tradition of being continuous rather than discrete, is another rich source of applications. Here, many of these fall into two categories: periodic phenomena and special functions. Here are just a few spatially or temporally periodic phenomena in physics: acoustics, diffraction, antenna design, dynamical systems, resonances of astronomical bodies, and crystal (and quasicrystal) structure. Number theory also has an affinity with special functions, for instance those arising in statistical mechanics. Here, the connection often involves the fact that many such special functions are the generating functions of sequences arising in additive (and sometimes multiplicative) number theory. It should be mentioned that there are many applications of number theory that do not come from computers or physics, for instance in chemistry, engineering, biology, and the arts. This book will introduce the reader to many of the above fields, giving some idea of the riches that exist. Many riches remain unmentioned, but the subject is too large to encompass in a single volume.

The title of the Short Course and of this book are a salute to the famous phrase of Wigner, who marveled at "the unreasonable effectiveness of mathematics" in the real world. Until recently, few people would have applied this phrase to number theory, but the authors and I hope that this volume will help to further the ferment that exists in the applications of number theory.

Stefan A. Burr

Proceedings of Symposia in Applied Mathematics
Volume 46, 1992

The Unreasonable Effectiveness of Number Theory
in Physics, Communication and Music

Manfred R. Schroeder

ABSTRACT. This paper samples applications of number theory in the real world, ranging from concert hall acoustics to general relativity.

1. Introduction

Number theory has been considered since time immemorial to be the very paradigm of "pure" (some would say useless) mathematics. According to Carl Friedrich Gauss, "mathematics is the queen of sciences – and number theory is the queen of mathematics." What could be more beautiful than a deep, satisfying relation between whole numbers. (One is almost tempted to call them *wholesome* numbers.) Indeed, it is hard to come up with a more appropriate designation than their learned name: the integers – meaning the "untouched ones." How high they rank, in the realms of pure thought and aesthetics, above their lesser brethren: the real and complex numbers – whose first names virtually exude unsavory involvement with the complex realities of everyday life!

Yet the theory of integers can provide totally unexpected answers to real-world problems. In fact, discrete mathematics is taking on an ever more important role. If nothing else, the advent of the digital computer and digital communication has seen to that. But even earlier, in physics, the emergence of quantum mechanics and discrete elementary particles put a premium on the methods and, indeed, the spirit of discrete mathematics.

1991 Mathematics Subject Classification. Primary 11A07, 11B83.

This paper is in final form and no version of it will be submitted for publication elsewhere.

© 1992 American Mathematical Society
0160-7634/92 $1.00 + $.25 per page

In mathematics proper, Hermann Minkowski, in the preface to his introductory book, on number theory, *Diophantische Approximationen;* published in 1907 (the year he gave special relativity its proper four-dimensional clothing in preparation for its journey into general covariance and cosmology) expressed his conviction that the "deepest interrelationships in analysis are of an arithmetical nature."

Yet much of our schooling concentrates on analysis and other branches of continuum mathematics to the virtual exclusion of number theory, group theory, combinatorics and graph theory. As an illustration, at a recent symposium on information theory, the author met several young mathematicians working in the field of primality testing, who – in all their studies up to the Ph.D. – had not heard a single lecture on number theory!

Or, to give an earlier example, when Werner Heisenberg discovered "matrix" mechanics in 1925, he did not know what a matrix was (Max Born had to tell him), and neither Heisenberg nor Born knew what to make of the appearance of matrices in the context of the atom. (David Hilbert is reported to have told them to go look for a differential equation with the same eigenvalues, if that would make them happier. They did not follow Hilbert's well-meant advice and thereby may have missed discovering the Schrödinger wave equation.)

Integers have repeatedly played a crucial role in the evolution of the natural sciences. Thus, in the 18th century, Lavoisier discovered that chemical compounds are composed of fixed proportions of their constituents which, when expressed in proper weights, correspond to the ratios of *small integers*. This was one of the strongest hints to the existence of atoms; but chemists, for a long time, ignored the evidence and continued to treat atoms as a conceptual convenience devoid of physical meaning. (Ironically, it was from the statistical laws of *large* numbers, in Einstein's and Smoluchowski's analysis of Brownian motion at the beginning of our own century, that the irrefutable reality of atoms and molecules finally emerged.)

In the analysis of optical spectra, certain integer relationships between the wavelengths of spectral lines emitted by excited atoms gave early clues to the structure of atoms, culminating in the creation of matrix mechanics in 1925, an important year in the growth of *integer physics*.

In 1882, Rayleigh discovered that the ratio of atomic weights of oxygen and hydrogen is not 16:1 but 15.882:1. These near-integer ratios of atomic weights suggested to physicists that the atomic nucleus must be made up of integer numbers of similar nucleons. The deviations from integer ratios later led to the discovery of elemental isotopes.

And finally, small divergencies in the atomic weight of pure isotopes from exact integers constituted an early confirmation of Einstein's famous equation $E = mc^2$, long before the "mass defects" implied by these integer discrepancies blew up into those widely noticed, infamous mushroom clouds.

On a more harmonious theme, the role of integer ratios in musical scales has been appreciated ever since Pythagoras first pointed out their importance. The occurrence of integers in biology – from plant morphology to the genetic code – is pervasive. It has even been hypothesized that the North American 17-year cicada selected its life cycle because 17 is a prime number, prime cycles offering better protection from predators than nonprime cycles. (The suggestion that the 17-year cicada "knows" that 17 is a *Fermat* prime has yet to be touted though.)

Another reason for the resurrection of the integers is the penetration of our lives achieved by that 20th-century descendant of the abacus, the digital computer. (Where did all the slide rules go? Ruled out of most significant places by the ubiquitous pocket calculator, they are sliding fast into restful oblivion.)

An equally important reason for the recent revival of the integer is the congruence of *congruential arithmetic* with numerous modern developments in the natural sciences and digital communication – especially "secure" communication by cryptographic systems. Last not least, the proper protection and security of computer systems and data files against computer "viruses" and other intrusions rest largely on keys ("digital signatures") based on congruence relationships.

In congruential arithmetic, what counts is not a numerical value per se, but rather its remainder or *residue* after division by a *modulus*. Similarly, in wave interference (be it of ripples on a lake or electromagnetic fields on a hologram plate) it is not path differences that determine the resulting interference pattern, but rather residues after dividing by the wavelength. For perfectly periodic events, there is no difference between a path difference of half a wavelength and one-and-a-half wavelengths: in either case the interference will be destructive.

One of the most dramatic consequences of congruential arithmetic is the existence of the chemical elements as we know them. In 1913, Niels Bohr postulated that certain integrals associated with electrons in "orbit" around the atomic nucleus should have integer values, a requirement that 10 years later became comprehensible as a wave interference phenomenon of the newly discovered de Broglie matter waves: In essence, integer-valued integrals meant that path differences are divisible by the electron's wavelength without leaving a remainder.

2. Music and numbers

Ever since Pythagoras, small integers and their ratios have played a fundamental role in the construction of musical scales. There are good reasons for this preponderance of small integers both in the production and perception of music. String instruments, as abundant in antiquity as today, produce simple frequency ratios when their strings are subdivided into equal lengths: shortening the string by one half produces the frequency ratio 2:1, the octave; and making it a third shorter produces the frequency ratio 3:2, the perfect fifth.

In perception, ratios of small integers avoid unpleasant beats between harmonics. Apart from the frequency ratio 1:1 ("unison"), the octave is the most easily perceived interval. Next in importance comes the perfect fifth. Unfortunately, as a consequence of the fundamental theorem of arithmetic, musical scales exactly congruent modulo the octave cannot be constructed from the fifth alone because there are no positive integers k and m such that

$$\left[\frac{3}{2}\right]^m = \left[\frac{2}{1}\right]^k .$$ (1)

However, there are good approximation to (1). Writing

$$3^m \cong 2^n$$

or

$$\log_2 3 \cong \frac{n}{m} ,$$

we see that we need a rational approximations to $\log_2 3$. The proper way of doing this is to expand $\log_2 3$ into a *continued fraction*

$$\log_2 3 = [1, 1, 1, 2, 2, \cdots]$$

which yields the close approximation m $=$ 12, n $=$ 19. In other words, if we want to make a good fifth with an equal-tempered (equal frequency ratio) scale, the basic interval $1:2^{1/12}$, the semitone, recommends itself. In fact, the semitone interval has come to dominate much of Western music. The equal-tempered fifth comes out as

$$2^{7/12} = 1.498 \cdots \tag{2}$$

Another fortunate number-theoretic coincidence is the fact that 7, the numerator in the exponent in (2), is coprime to 12. As a consequence we can reach all 12 notes of the octave interval by repeating the fifth (modulo the octave). This is the famous Circle of Fifths.

Of course, (2) is not an *exact* equality and some compromises have to be made in the construction of scales. Pythagoras used the perfect fifth (3:2) and the perfect fourth (4:3), but fudged on the minor and major thirds, which come out as 1:1.185 and 1:1.265, rather that 6:5 and 5:4, respectively. (How Pythagoras must have wished the fundamental theorem out of existence!)

More recently J. R. Pierce has tried to go Pythagoras one better by constructing a musical scale for the "tritave" (the frequency interval 3:1) based on the integer ratio 5:3. Pierce used trial and error, but we simply expand $\log_3 5$ into a continued fraction. This yields the close approximation $5 \cong 3^{19/13}$, the tritave should be subdivided into 13 equal intervals. (As a bonus – a number-theoretic fluke – $3^{1/13}$ also allows a very close approximation of the next prime 7. To wit: $7 \cong 3^{23/13}$.)

3. Concert halls and quadratic residues

There is another connection between music and numbers: concert hall acoustics. Extensive physical tests and psychophysical evaluation of the acoustic qualities of concert halls around the world have established the importance of *laterally* traveling sound waves. (Such waves produce dissimilar signals of a listener's two ears, a kind of stereophonic condition that is widely preferred for music listening.)

In order to convert sound waves traveling longitudinally (from the stage via the ceiling to the main listening areas) into lateral waves, the author has recommended ceiling structures that scatter sound waves, without absorption, into broad lateral patterns. In the physicist's language, concert hall ceilings (and perhaps other surfaces too) should be *reflection phase-gratings* with equal energies going into the different diffraction orders (the different lateral directions).

How should one go about designing such an ideal scatterer for sound (or light or radar waves)? Curiously, one answer comes from a classical branch of number theory that has exercised the great Gauss for a long time: quadratic residues. Consider a surface structure whose reflection coefficient r_n varies in equidistant steps along one axis according to

$$r_n = \exp(2\pi i n^2/p) , \quad n = 0 , \pm 1, \pm 2, \cdots \qquad (3)$$

where p is a prime and n^2 may be replaced by $(n^2) \bmod p$, its least nonnegative residue modulo p.

It is easy to show that the discrete Fourier Transform (DFT) of r_n has constant magnitude. As a physical consequence, the intensities of the wavelets scattered into different directions from a surface with reflection coefficients (3) will have equal magnitudes (in the customary Kirchhoff approximation of diffraction theory [1]). The scattering angles α_k are given by the wavelength λ and the step size w (corresponding to $\Delta n = 1$):

$$\sin \alpha_k = \frac{k\lambda}{pw} , \qquad |k| \le \frac{pw}{\lambda} , \qquad (4)$$

yielding $2 \lfloor pw/\lambda \rfloor + 1$ different angles. (The "Gauss brackets" $\lfloor \ \rfloor$ stand for rounding down to the nearest integer.) The different reflection coefficients are realized by "wells" of different depths

$$d_n = \frac{\lambda}{2p} (n^2) \bmod p ,$$

as illustrated in Fig. 1 for $p = 17$. Such wells give a round-trip phase change of $2d_n \cdot 2\pi/\lambda$ in accordance with the phase requirement of (3). In (4), λ is the *longest* wavelength to be scattered. For any integral submultiple of that wavelength, λ/m, the reflection coefficients (3) are changed to r_n^m, which for $m \not\equiv 0 \pmod p$ has the same flat Fourier property as (3). For w one chooses typically half the smallest wavelength to be scattered over $\pm\pi/2$. Figure 2 shows the diffraction pattern of the grating in Fig. 1 for one third the longest wavelength.

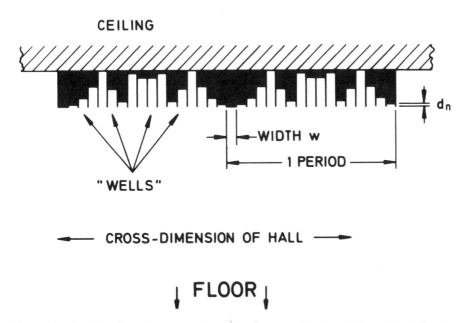

Figure 1. A reflection phase-grating, based on quadratic residues, for effective scattering of waves. In this example residues are taken modulo p = 17. Such structures are useful in radar and sonar the diffusion of coherent light, noise abatement – and concert hall acoustics.

INCIDENT WAVE ⇓

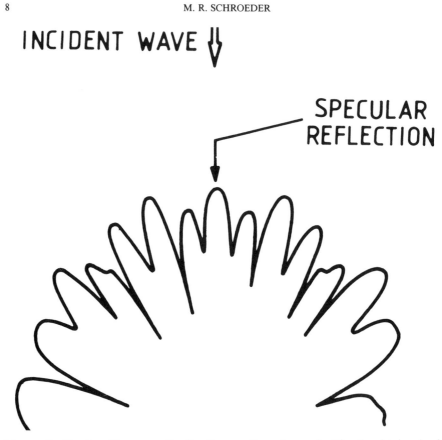

SPECULAR
REFLECTION

Figure 2. Scatter diagram of reflection grating shown in Fig. 1, obtained with sound waves at three times the "design" frequency. With p = 17, the grating scatters effectively over a frequency range of 1 to p − 1 = 16 (four musical octaves).

4. Wave diffraction and primitive roots

In some applications, including concert hall acoustics, it may be desirable to attenuate the specular reflection (the "zeroeth diffraction order") relative to the other diffraction orders. Are such "super scattering" phase gratings possible? The answer is yes and it comes, again, from number theory.

Instead of quadratic residues the author [2] has suggested powers of primitive roots to construct reflection coefficients. Specifically (3) is replaced by

$$r_n = e^{2\pi i g^n/p} , \qquad (5)$$

where g is a primitive root of the prime p (g is a primitive root of p if its powers, g^n, generate all the $p - 1$ different nonzero integers modulo p). The sequence r_n is then periodic with period $p - 1$. It is not difficult to show that the Discrete Fourier Transform R_k of (5) has constant magnitude except for $|R_0|$ which is p times smaller than $|R_k|$ for $k \not\equiv 0 \pmod{p - 1}$.

Figure 3 shows a scatter diagram of a microwave reflection phase grating based on $p = 7$ and $g = 3$. Note the attenuated zeroeth diffraction order between the six strong "lobes" corresponding to $k = \pm1, \pm2, \pm3$.

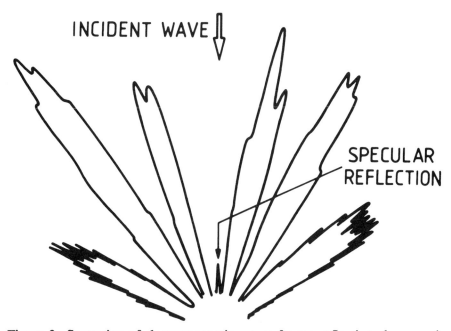

INCIDENT WAVE

SPECULAR REFLECTION

Figure 3. Scattering of electromagnetic waves from a reflection phase-grating based on powers of a primitive root g of a prime p. (Here $p = 7$ and $g = 3$.) Note strong reflections in the six non-specular directions and the weak specular reflection in the center between them.

The success of phase gratings based on primitive roots raises the following mathematical question: The least positive residues of g^n modulo p generate a permutation of the integers 1 to $p - 1$. The $\phi(p - 1)$ primitive roots of p generate $\phi(p - 1)$ different permutations, which is typically a small fraction of all possible permutations. Not counting cyclical shifts and reflections, there are $(p - 2)!/2$ "inequivalent" permutations. For $p = 17$, we have thus 653 837 184 000 such permutations, of which $\phi(16) = 8$ (about one in 10^{11}) have the flat Fourier property. Questions: are there other permutations that, when used instead of g^n in (5), have this property?

5. Forbidding property of the Fermat primes

Since Gauss's great discovery of 30 March 1796 concerning the factorization of $x^p - 1$, where p is a Fermat prime,

$$p = 2^{2n} + 1 ,$$

these primes have enjoyed great esteem, spreading from the factoring factories of the experts to the sweatshops of the amateur. The "geometrical" constructions of regular p-gons rests on the fact that $p - 1$ has only one prime factor, namely 2. Now, in the context of wave diffraction, this same property – instead of permitting a particular construction – *forbids* a desirable application, namely the construction of two (or higher) dimensional phase arrays with the flat Fourier property mentioned above.

The two-dimensional generalization of the periodic sequence (5) is the periodic array $r_{k\ell}$ with

$$k \equiv (n)_{\text{mod } K} \quad \text{and} \quad \ell \equiv (n)_{\text{mod } L} ,$$

where K and L (in best Chinese remainder fashion) are two coprime factors of $p - 1$. Thus there are no n-dimensional $(n > 1)$ phase gratings based on the known Fermat primes 3, 5, 17, 257, 65 537. (The smallest prime for a three-dimensional array is 31, having a $2 \cdot 3 \cdot 5$ "unit cell.")

6. Euler totients and cryptography

One of the most spectacular applications of number theory in recent times is *public-key cryptography* in which each potential recipient of a secret message *publishes* his encryption key, thereby avoiding the (often substantial) problems of secure secret-key distribution. But how can a key be public and yet produce secret messages? The answer is based on Euler's totient function $\phi(r)$ and the role it plays in inverting modular exponentiation. The public key consists of a modulus r and an exponent s, coprime to $\phi(r)$. The message is represented by an integer M, $1 < M < r$, and the encrypted message E is given by a number in the same range, calculated as follows

$$E \equiv M^s \bmod r .\tag{6}$$

Decrypting E is accomplished by calculating $E^t \bmod r$, where the decrypting exponent t is given by

$$ts \equiv 1 \bmod \phi(r) ,\tag{7}$$

i.e., $ts = k\phi(r) + 1$ for some k. With such a t,

$$E^t = M^{st} \equiv M^{k\phi(r)+1} \bmod r .$$

which, according to Euler's theorem, give the message M back.

So far, so good and – theoretically – trivial. The trick in public-key encryption is to choose r as the product of two very large primes, each say 200 digits long. (There is no paucity of such primes, and enough for all foreseeable purposes can be easily ferreted out from the jungle of composites in the 10^{200} neighborhood.)

Now, with a composite r, prescription (7), so easily written down, becomes practically impossible to apply because $\phi(r)$ can be calculated only if the factors of r are known – and this knowledge is *not* published. In modern parlance, the mapping (6) is a *trap-door function*.

A trap-door function is, as the name implies, a (mathematical) function that is easy to calculate in one direction but very hard to calculate in the opposite direction. For example, it takes a modern computer only microseconds to multiply two 100-digit numbers. By contrast, to decompose a 200-digit number, having two 100-digit factors, into its factors can take "forever," even on the fastest computers available in 1992 and using the most efficient factoring algorithms known today [3].

On the other hand, knowing the factors of r, as the legitimate recipient of the encrypted message E does, $\phi(r)$ can be easily calculated and decrypting becomes possible. The decrypting exponent t is obtained by Euclid's algorithm [7] or by solving (7) directly:

$$t \equiv s^{\phi(\phi(r))-1} \bmod \phi(r) .$$

Not so long ago, the most efficient factoring algorithms on a very fast computer were estimated to take trillions of years. But algorithms get more efficient by the month and computers become faster and faster every year, and there is no *guarantee* that one day a so-called "polynomial-time" algorithm will not emerge that will allow fast factoring of even 1000-digit numbers. Few mathematicians believe that a true polynomial-time algorithm is just around the corner, but there also seems to be little prospect of proving that this will never occur. This is the Achilles heel of public key cryptography.

7. Sequences from finite fields

Certain periodic sequences with elements from the finite (Galois) field GF(p), formed with the help of irreducible polynomials over GF(p^m), have unique and much sought-after correlation and Fourier transform properties [7]. These Galois sequences, as I have called them, have found ingenious applications in error-correcting codes (compact discs and picture transmissions from inter-planetary satellites) and in precision measurements from physiology to general relativity. Other applications are in radar and sonar camouflage, because Galois sequences, like quadratic residues (see above), permit the design of surfaces that scatter incoming waves very broadly, thereby making the reflected energy "invisible" or "inaudible." A similar application occurs in work with coherent light, where a "roughening" of wavefronts (phase randomization) is often desired (for example, to avoid "speckles" in holograms). Light diffusors whose design is based on Galois arrays are in a sense the ultimate in frosted (milk) glass. Finally, Galois sequences allow the design of loudspeaker and antenna arrays with very broad radiation characteristics [2].

8. Error correction codes from Galois fields

Galois sequences, with periods n = p^m − 1, are constructed with the help of an irreducible polynomial g(x) of degree m with coefficients from GF(p), such that g(x) is a factor of x^n − 1 but not a factor of x^r − 1 for r < n.

Binary Galois sequences (p = 2) with elements 0 and 1 (or, in certain other applications, 1 and −1) are the most important practical case. For p = 2 and m = 4, an irreducible polynomial with the stated property is

$$g(x) = 1 + x + x^4 ,$$

from which the recursion

$$a_{k+4} = a_{k+1} + a_k \qquad (8)$$

is obtained. Beginning with the initial condition 1000 (or almost any other tuple, except the all-zero tuple) (8) generates the binary Galois sequence of periodic of period length $2^4 - 1 = 15$:

1000, 10011010111; etc. (repeated periodically) .

The error correcting properties of codes based on such sequences result from the fact that the $2^4 = 16$ different initial conditions generate 16 different code words of length 15, that form a "linear code" (the sum of two code words is another code word). These code words define therefore a 4-dimensional linear subspace of the 15-dimensional space with coordinates 0 and 1. In fact, the 16 code words describe a simplex (in 3 dimensions a simplex is a tetrahedron) in that space and the resulting code is therefore called a *Simplex Code* [4].

Its outstanding property is that every pair of code words has the same Hamming distance (the number of 0,1 disparities), namely $2^{m-1} = 8$. Thus, the code can recognize up to $2^{m-2} = 4$ errors and correct up to $2^{m-2} - 1$ errors. The price for this error correcting property is a reduced signaling efficiency, namely $m/n = 4/15$, where $n = 2^m - 1$ is the length of the code word.

Several other codes can be derived from the simple Simplex Code. For example, the famous (and historically early) *Hamming Codes*. The code words of a Hamming Code are given by the orthogonal subspace of the Simplex Code of the same length. Hamming codes carry $n - m$ information bits and m check bits, and can correct precisely one error. The functioning of the Hamming Code for $m = 3, n = 7$, is illustrated in Fig. 4.

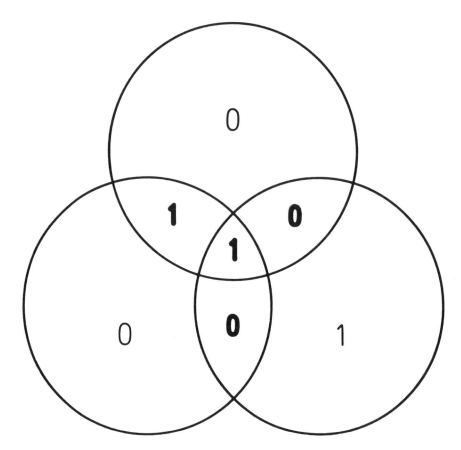

Figure 4. Venn diagram to illustrate the single-error correcting property of the original Hamming code having 4 information bits (fat 0s and 1s) and 3 check bits (thin 0s and 1s). The parity in each of the 3 circles must be even. A single error, which entails one or several odd parities, can be uniquely detected and corrected.

The $n - m = 4$ information bits, say 1001, are entered into the 4 inner areas of the Venn diagram (indicated by fat characters in Fig. 4). The $m = 3$ check bits (thin characters) are entered into the 3 outer areas such that the parity in each circle is even (the sum modulo 2 equals 0).

The receiver of a code word, which may have been contaminated in transmission, checks the parity in each circle and marks all circles with odd parity. The intersection of these circles then specifies uniquely a single bit error (including in the check bits themselves). These 3 parity checks allow the receiver to distinguish between precisely $2^3 = 8$ different possibilities: a single error in any of the 7 transmitted bits or *no* error. No wonder the Hamming Code is called a perfect code.

9. Correlation and fourier properties of Galois sequences

For many purposes it is advantageous to use the elements $s_k = 1$ or -1 instead of $a_k = 0$ or 1. The mapping is

$$s_k = 2a_k - 1 .$$

Using this notation, the constant Hamming distance between code words of a Simplex Code (whose members are generated by cyclic shifts) translates immediately into the following circular auto-correlation property

$$c_r: = \sum_{k=1}^{n} s_k s_{k+r} = -1 \quad \text{for } r \not\equiv 0 \bmod n ,$$

and, of course, $c_r = n$ for $r \equiv 0 \bmod n$. As a result of this two-valuedness of c_r, the Fourier transform of s_k:

$$S_m = \sum_{k=1}^{n} s_k \exp(-i2\pi km/n)$$

has constant magnitude for $m \not\equiv 0 \bmod n$. In the lingo of the physicist and computer scientist: the sequence s_k has a flat (or "white") power spectrum.

If we identify the index k with (discrete) physical time, then we can say that the "energy" $|s_k|^2 = 1$, of the sequence s_k is equally distributed over all *time* epochs. And because of $|S_m|^2$ is constant, we can make the same claim with respect to the distribution of energy over all (non-zero) *frequency* components. This equal "energy spreading" of the Galois sequences s_k with period length $n = 2^m - 1$, obtained with the help of polynomials over $GF(p^m)$, has many impressive applications, some of the more astounding ones occurring in the interplanetary distance measurements.

10. Galois sequences and the fourth effect of general relativity

General Relativity, the theory of gravitation propounded by Einstein in November 1915 in Berlin (and 5 days earlier by Hilbert in Göttingen) passed three important experimental tests during Einstein's lifetime:

1) The perihelion motion of the orbit of the planet Mercury, which was already known from earlier astronomical observations.

2) The bending of light waves near the sun, first observed during the total eclipse of 1919 by the Eddington expedition.

3) The gravitational red shift, first seen in the light from massive stars, but now measurable even on Earth herself using the ultrasensitive Moessbauer effect.

A fourth effect inherent in Einstein's theory was not confirmed until fairly recently: the slowing of electro-magnetic radiation in a gravitational field.[1] This effect was observed by means of radar echoes from the planets Venus and Mercury as they disappeared behind the sun as seen from the earth ("superior conjunction"). In that position, both the outgoing and returning radar waves have to travel near (indeed around) the sun. Even after taking plasma effects near the sun's surface and other factors into account, physicists found an extra delay of 200 μs – very close to the prediction of general relativity [5].

Why was this measurement not done long ago? The reason is that the echo energy from Mercury – exceedingly weak even when visible – drops to 10^{-27} of the outgoing energy as the planet slips behind the sun. The astounding fact that reliable results have been obtained in spite of these miniscule reflected energies is due mainly to the proper choice of the transmitted sequence of radar pulses, based on primitive polynomials over finite number fields.

---

[1] In the long struggle to put his principle of general equivalence of different references frames into proper mathematical clothing, Einstein discovered – as early as 1909 – that the speed of light could not be constant (as in special relativity) but must depend on the gravitational potential ϕ. Although he had no general theory then, Einstein found that, to first order, $c(\phi) = c + \phi/c$, where c is the usual vacuum velocity of light in field-free space (Note: $\phi \leq 0$). Ironically, the slowing of radiation i n gravitational fields, although appreciated very early, was not considered a testable proposition until the perfection of radar technology, using Galois sequences, in the second half of this century. The reason for this delay in testing the extra delay was, of course, that no one could picture himself (or anyone else, for that matter) floating next to the sun, stopwatch in hand, clocking the passing photons.

11. Chinese remainders feed fast algorithm

In the computation of discrete circular convolutions

$$h(n) = \sum_{k=1}^{M} f(k)\,g(n-k)_{\text{mod } M}, \; n = 1,2, \ldots M , \tag{9}$$

a pervasive task in numerous numerical applications [6], the number of arithmetical "operations" (multiplications and additions) is M^2, where M is the period of the involved sequences. It is easy to show [7] that if M has $r > 1$ coprime factors

$$M = m_1 m_2 \ldots m_r ,$$

then the one-dimensional convolution (9) can be converted into an r-dimensional convolution by expressing the summating index n in the *Sino notation:*

$$n = \sum_i n_i N_i M/m_i \bmod M ,$$

where n_i is least positive remainders of n modulo m_i, and N_i is given by the congruence

$$N_i M/m_i \equiv 1 \bmod m_i$$

The necessary summations over the n_i require a total of $M \Sigma m_i$ operations, which can be considerably smaller than M^2. For example, for $M = 1\,007\,760 = 13 \cdot 15 \cdot 16 \cdot 17 \cdot 19$, the number of operations drops by a factor $M/\Sigma m_i = 12\,597$ – a very substantial saving, comparable to the economy offered by the Fast Fourier Transform (FFT), which also converts a one-dimensional operation into a multi-dimensional one. In fact, the fast Chinese convolution [7] described here complements nicely the FFT which works most efficiently if all factors m_i of M equal 2 rather than being coprime.

12. Epilogue

It is clear that only a sprinkling of the numerous applications of number theory outside mathematics proper could be mentioned here. Among the many topics that had to be counted out from this brief account are

1. The application of continued fractions to electrical network problems which, incredible, led to the construction of the "squared square" [7] – long considered impossible [8]. (The squared square is a square with integer sides, completely covered, without overlap, by smaller incongruent integer squares.

2. Heat conduction in thin tours: the solution is based on the representation of integers by the sum of two squares [9].

3. The eigenvalue distribution of normal modes in cubical (and near cubical) resonators, which depends on the representation of integers by the sum of three squares [7]. (The author's first encounter with number theory, in his Ph.D. thesis on concert hall acoustics.)

4. Search algorithms, game strategies [10] and countless other applications based on Fibonacci numbers.

5. Certain unexpected properties of the zeroes of Riemann's zeta-function, found by A. Odlyzko [11], and their possible relation to the Wigner distribution function (which governs the distribution of energy levels in the atomic nucleus and eigenfrequencies in complex vibrational systems).

6. And, most recently, the elucidation of the structure of quasi-crystals, a new state of matter combining "forbidden" 5-fold rotational symmetry and sharp, crystal-like, diffraction patterns [12].

What riddle will be solved next by number theory? Is this effectiveness of the higher arithmetic completely unreasonable? Or are we witnessing here a "pre-established harmony" à la Leibniz between mathematics and the real world?

REFERENCES

[1]   M. R. Schroeder, "Diffuse sound reflection by maximum-length sequences," J. Acoust. Am. vol. 57, 140-150 (1975). See also: M. R. Schroeder "Toward better acoustics for concert halls," *Physics Today*, 24-30, Oct. 1980.

[2]   M. R. Schroeder, "Constant-amplitude antenna arrays with beam patterns whose lobes have equal magnitudes," *Archiv für Electronic und Uebertragungstechnik (Electronics and Communication)* vol. 34, 165-168 (1980).

[3]   C. Pomerance, "Recent developments in primality testing," The Mathematical Intelligencer, vol. 34, 97-105 (1981).

[4]   F. J. MacWilliams and N. J. Sloane, *The Theory of Error-Correcting Codes,* North-Holland, Amsterdam 1978.

[5]   I. I. Shapiro et al., "Fourth test of general relativity," *Phys. Rev. Lett.,* vol. 20, 1265-1269 (1968).

[6]   J. H. McClelland and C. M. Rader, *Number Theory in Digital Signal Processing,* Prentice-Hall, Englewood Cliffs 1979.

[7]   M. R. Schroeder, *Number Theory in Science and Communication –* With Applications in Cryptography Physics, Digital Information, Computing and Self-similarity, Second Enlarged Edition, Springer-Verlag, Berlin 1986.

[8]   C. J. Bouwkamp, A. J. Duijvestijn and P. Medema, *Tables Relating to Simples Squared Rectangles,* Dept. of Mathematics and Mechanics, Technische Hogeschool, Eindhoven 1960.

[9]   J. Rohlfs, *Mathematische Miniaturen,* in F. Hirzebruch (ed.), Birkhäuser, Basel, 75-91 (1983).

[10] E. R. Berlekamp, H. H. Conway, and R. K. Guy, *Winning Ways,* Academic Press, London 1981.

[11] A. M. Odlyzko, "On the Distribution of Spacing Between Zeroes of the Zeta Function," *Math. Comp.,* vol. 48, 273-308, Jan. 1987.

[12] M. R. Schroeder, *Fractals, Chaos, Power Laws: Minutes from an Infinite Paradise,* W. H. Freman, New York 1991.

Drittes Physikalisches Institut, University of Göttingen, W-3400 Göttingen, Germany, and AT&T Bell Laboratories (ret.)

Proceedings of Symposia in Applied Mathematics
Volume 46, 1992

The Reasonable and Unreasonable Effectiveness of Number Theory in Statistical Mechanics

GEORGE E. ANDREWS

1. Introduction.

This paper will be somewhat more restricted than the title implies. We shall concern ourselves only with additive number theory and how it relates to statistical mechanics. Also it must be noted at the onset that this paper was presented in a session entitled The Unreasonable Effectiveness of Number Theory. The course description suggests that thirty years ago all applications of number theory would have seemed unreasonable. However that is really not the case in statistical mechanics. In Section 2, I shall try to sketch something of the "reasonable" effectiveness of number theory by describing several topics (dating back to the 1930's and '40's) where the theory of partitions seems to arise naturally. Toward the end of that section we shall see small surprises. However in Section 3 we shall discuss R. J. Baxter's solution of the Hard Hexagon Model and how the Rogers-Ramanujan identities fit in. This is indeed starting to be unreasonable. We pass on to the truly bizarre in Section 4 where we describe how intrinsically outrageous algebraic identities found by Ramanujan for the Rogers-Ramanujan functions actually play a perfect part in illuminating further aspects of the Hard Hexagon Model. In Section 5 we look at Baxter's beautiful and not widely known general theorem on modular relations for Rogers-Ramanujan type functions. We conclude with brief reference to further developments.

2. Early Applications.

We begin by noting that any time the number of ways of writing a number as a sum of other numbers arises the theory of partitions can't be far off. We begin by quoting from a 1937 paper by C. Van Lier and G. E. Uhlenbeck [20]:

1991 *Mathematics Subject Classification.* Primary 82A68, 11P68 Secondary 82-03, 05A19.
Partially supported by National Science Foundation Grant DMS-8202695-03
This paper is in final form and no version of it will be submitted for publication elsewhere.

© 1992 American Mathematical Society
0160-7634/92 $1.00 + $.25 per page

"Let ϵ_i be the individual energy levels, and $n_i (= 0$ or $1)$ the number of particles with energy ϵ_i. Write:

$$N = \sum_i n_i \qquad E = \sum_i \epsilon_i n_i$$

for the total number of particles resp. total energy. The density $\rho(E, N)$ of the energy levels of the whole gas around the energy will be given by the number of ways in which the state E can be realized.

This will be given by the coefficient of $x^N y^E$ in the development:

$$\sum_{n_1=0}^{1} \sum_{n_2=0}^{1} \cdots x^{\sum n_i} y^{\sum \epsilon_i n_i} = \prod_i (1 + xy^{\epsilon_i}).$$ "

(N. Bohr and F. Kalckar [8] made similar observations at essentially the same time).

From here the rest of their paper consists of asymptotic analysis of this and related generating functions, interpretation of the results in physics and generalization. The above paragraph makes clear at the outset that the methods of additive analytic number theory will guide the physicists in what they can hope to discover. I would call this application quite reasonable; the problem under consideration is naturally formulated in terms of the theory of partitions. Precisely the same observations may be made in many other statistical applications in and outside of physics. For example, the famous paper [13] by Mann and Whitney wherein they provide the full elucidation of the Wilcoxon test in non-parametric statistics is precisely a study of a distribution defined by partitions (masquerading as inversions in sequence of 0's and 1's [12]) in which the number of parts and their size are explicitly restricted. In 1946, Auluck and Kothari [5] considered several "thermodynamical assemblies corresponding to the partition functions familiar in the theory of numbers."

A slightly less immediately direct application was pursued by Temperley [19] in his work on the form of crystal structures. In the classical theory of partitions developed by Sylvester [15], each partition, for example $7+5+5+5+4+2+1+1$ has associated with it a Ferrers graph:

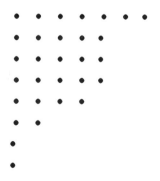

wherein each row represents a part of the partition. In the first of Temperley's examples [19; **p. 684–685**] the above Ferrers graph would be viewed as an array of 30 molecules forming a crystal between two walls at right angles. The number of such arrays with n molecules is $p(n)$, the total number of partitions of n, and Temperley is then able to apply the known asymptotics of $p(n)$ to deduce various physical parameters associated with his model. Temperley uses the same exact setting in three-dimensions and here he involves MacMahon's work on plane partitions [11] together with E. M. Wright's subsequent asymptotics for the plane partition function [22]. All of these models are quite straightforward and nothing more subtle than the Ferrers graph is involved in the creation of the model. Furthermore while Temperley does rely on the amazing asymptotic Hardy-Ramanujan-Rademacher series for $p(n)$ [**1; Ch. 5**], it is nonetheless the case that this is unnecessary, and indeed the standard saddle-point method of asymtotic analysis is adequate for the results that Temperley requires. No number theory magic has been seen so far. However it would be inaccurate to suggest that none appears in Temperley's work. In Section 3 of [**19**], he considers: "The growth of a crystal on a plane substrate." In this case, we are examining "two-sided" Ferrers graphs. The example Temperley uses is

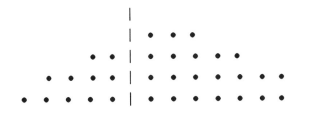

This (by means of the dotted line) may be split into two partitions; the right-hand one is $7 + 7 + 5 + 3$ and the left-hand one is $5 + 4 + 2$. If such an array consists of n rows of dots, the generating function for partitions on the right of the dotted line is

$$\frac{q^n}{(1 - q)(1 - q^2) \ldots (1 - q^n)},$$

while the generating function for the partitions on the left is

$$\frac{1}{(1 - q)(1 - q^2) \ldots (1 - q^{n-1})}.$$

Consequently, the full generating function is formed by multiplying together the above two functions and summing over all positive n:

$$(2.1) \qquad f(q) = \sum_{n=1}^{\infty} \frac{q^n}{(1 - q)^2 (1 - q^2)^2 \cdots (1 - q^{n-1})^2 (1 - q^n)}$$

Now a genuine problem arises for treating $f(q)$. The series does not converge rapidly enough to allow reasonable asymptotic treatment of the coefficient. However by a neat argument Temperley transforms $f(q)$ into:

$$(2.2) \qquad f(q) = \frac{q - q^3 + q^6 - q^{10} + q^{15} - q^{21} + \cdots}{\prod\limits_{n=1}^{\infty} (1 - q^n)^2},$$

a result that follows immediately from setting $a = b = 0$, $c = q^2$, $t = q$ in Heine's transformation [1; p. 19]:

$$(2.3) \qquad \sum_{n=0}^{\infty} \frac{(a)_n (b)_n t^n}{(q)_n (c)_n} = \frac{(b)_\infty (at)_\infty}{(c)_\infty (t)_\infty} \sum_{n=0}^{\infty} \frac{(c/b)_n (t)_n b^n}{(q)_n (at)_n},$$

where

$$(2.4) \qquad (a)_n = (1 - a)(1 - aq) \cdots (1 - aq^{n-1}),$$

and

$$(2.5) \qquad (a)_\infty = \prod_{j=0}^{\infty} (1 - aq^j).$$

Identity (2.1) is quite adequate for yielding the necessary asymptotic information.

In summary then we have seen in this section how the idea of partition of numbers fits nicely and reasonably directly into several models from statistical mechanics. Only in passing from (2.1) to (2.2) do we get any hint that any of the identities arising in additive number theory may be of use. In the next section we shall see amazing identities arise inexorably in Baxter's solution of the Hard Hexagon model.

3. The Hard Hexagon Model.

The story of Baxter's solution of the Hard Hexagon model is both amazing and difficult. A full account is provided in Chapter 14 of Baxter's book [6]. However this chapter cannot be read independently from the preceding thirteen; the necessary method of corner transfer matrices (Chapter 13) evolves from the row-to-row transfer matrix method (Chapters 7-10) and the use of elliptic functions to parametrize solutions of the star-triangle relations is laid out in detail in Chapter 8. Therefore the material in this section will appear sketchy at best.

The Hard Hexagon model is a two-dimensional model. This means we are to consider the integer points in the Euclidean plane. We restrict oruselves initially to a large square centered on the origin. The integer points therein are numbered from 1 to N. The $i^{\underline{th}}$ integer point is assigned an occupancy number $\sigma_i = 0$ or 1 which tells us whether there is a particle occupying this point ($\sigma_i = 1$) or not ($\sigma_i = 0$). For the Hard Hexagon model we assume that if a given integer point is occupied then a particular six of the nearest eight integer points (see Figure 1) must be unoccupied as indicated.

FIGURE 1

Obviously there are many admissible ways of assigning values to the σ_i of the N points of our system. Each such assignment is called a state of the system producing a related energy $E(s)$ and a number of particles $n(s)$.

The underlying philosophy is that one can obtain information about some sort of atomic structure (the distribution of a film of liquid helium on a graphite plate is the physical structure underlying the Hard Hexagon model) by determining averages over all possible states of a large admissable system of particles (in the final analysis $N \to \infty$).

Gibbs defined the probability of a system being in state s by

$$(3.1) \qquad Z^{-1}e^{-\beta E(s)-\mu n(s)},$$

and assuming that the probabilities add up to 1, we see that

$$(3.2) \qquad Z = \sum_{\substack{\text{all states} \\ s}} e^{-\beta E(s)-\mu n(s)}.$$

Z is called the partition function (not to be confused with $p(n)$ and the partitions of integers mentioned earlier). In addition to Z itself, various other averages can be computed. For our discussion here we mention two. First $< n >$ the average number of particles

$$(3.3) \qquad < n >= \frac{1}{Z} \sum_{\substack{\text{all states} \\ s}} n(s)e^{-\beta E(s)-\mu n(s)},$$

and also $< \sigma_1 >$ the probability that a particle deep inside the system (e.g. the origin) is occupied;

$$(3.4) \qquad < \sigma_1 >= \frac{1}{Z} \sum_{\substack{\text{all states} \\ s}} \sigma_1 e^{-\beta E(s)-\mu n(s)}.$$

Now we assume that the only nonnegligible contributions to $E(s)$ occur due to interactions of particles around the unit squares (see Figure 2).

FIGURE 2

More succinctly

$$(3.5) \qquad E(s) = \sum_{\substack{\text{all faces} \\ i,j,k,\ell}} \epsilon(\sigma_i, \sigma_j, \sigma_k, \sigma_\ell)$$

We can now translate the additive nature of E and n back into our formula for Z to see that there exists ω such that

$$(3.6) \qquad Z = \sum_{\substack{\text{all} \\ \text{states}}} \prod_{\substack{\text{all} \\ \text{faces}}} \omega \begin{pmatrix} \sigma_\ell & \sigma_k \\ \sigma_i & \sigma_j \end{pmatrix}.$$

Through the magic of the corner transfer matrix method, Baxter was able to evaluate explicitly for the Hard Hexagon model Z and the related averages in terms of parameters of the system. Indeed Baxter is able to solve a generalized Hard Hexagon model wherein

$$(3.7) \qquad \omega \begin{pmatrix} d & c \\ a & b \end{pmatrix} = z^{(a+b+c+d)/4} e^{Lac + Mbd}$$
$$\times (1 - ab)(1 - bc)(1 - cd)(1 - da)$$

provided that several equations including

$$(3.8) \qquad z = (1 - e^{-L})(1 - e^{-M})/(e^{L+M} - e^L - e^M)$$

are satisfied, and this is done by parametrizing the solutions through the use of elliptic functions.

Historically Onsager was apparently the first to use elliptic functions in such problems; he employed them in his solution of the Ising model. Baxter provides a full account of the solution of the Ising model in [**6, Ch. 6**].

In the final analysis (after $N \to \infty$) Z and $< \sigma_1 >$ are obtained as

$$(3.9) \qquad Z = \text{Trace } ABCD$$

$$(3.10) \qquad < \sigma_1 > = \frac{\text{Trace } SABCD}{Z}$$

where A, B, C, D are the infinite commuting corner transfer matrices and S is the diagonal matrix with σ_1 on the main diagonal. Thus the whole problem has been boiled down to the computation of the eigenvalues of $ABCD$ which Baxter finds to be

$$(3.11) \qquad r_0^{2\sigma_1} q^{\sigma_2 + 2\sigma_3 + 3\sigma_4 + \cdots}$$

where the sequences $\{\sigma_i\}_{i=1}^{\infty}$ are now sequences of 0's and 1's with $\sigma_i + \sigma_{i+1} \leqq 1$, and

(3.12)
$$q = x^6$$

(3.13)
$$r_0^2 = \frac{-xG(x)}{H(x)}$$

(3.14)
$$G(x) = \prod_{n=1}^{\infty} \frac{1}{(1 - x^{5n-4})(1 - x^{5n-1})}$$

(3.15)
$$H(x) = \prod_{n=1}^{\infty} \frac{1}{(1 - x^{5n-3})(1 - x^{5n-2})}$$

with x a real parameter of the system lying in the interval $(-1, 0)$.

Consequently

(3.16)
$$Z = \sum_{\sigma_1, \sigma_2,} r_0^{2\sigma_1} q^{\sigma_2 + 2\sigma_3 + 3\sigma_4 + \cdots}$$
$$= \sum_{\substack{\sigma_2, \sigma_3, \cdots \\ \sigma_1 = 0}} q^{\sigma_2 + 2\sigma_3 + 3\sigma_4 + \cdots} + r_0^2 \sum_{\substack{\sigma_2, \sigma_3, \cdots \\ \sigma_1 = 1}} q^{\sigma_2 + 2\sigma_3 + 3\sigma_4 + \cdots}$$
$$\equiv F(0) + r_0^2 F(1),$$

while

(3.17)
$$< \sigma_1 > = \frac{1}{Z} \sum_{\sigma_1, \sigma_2, \cdots} \sigma_1 r_0^{2\sigma_1} q^{\sigma_2 + 2\sigma_3 + 3\sigma_4 + \cdots}$$
$$= \frac{r_0^2 F(1)}{F(0) + r_0^2 F(1)}.$$

It is now a straightforward matter to evaluate $F(0)$ and $F(1)$ in series much like the one we found for $f(q)$ in Section 2. Indeed if we define as did Baxter

(3.18)
$$F_m(\sigma_m) = \sum_{\sigma_{m+1}, \sigma_{m+2}, \cdots} q^{m\sigma_{m+1} + (m+1)\sigma_{m+2} + \cdots}$$

(keeping in mind the condition $\sigma_i + \sigma_{i+1} \leq 1$), then immediately

(3.19)
$$F_m(0) = F_{m+1}(0) + q^m F_{m+1}(1),$$

(3.20)
$$F_m(1) = F_{m+1}(0),$$

and if we expand F_m in a series of the form

(3.21)
$$F_m(\sigma_m) = \sum_{n=0}^{\infty} q^{nm} a_n(\sigma_m),$$

then we find by coefficient comparison in the functional equations that

(3.22)
$$F_m(\sigma_m) = \sum_{n=0}^{\infty} \frac{q^{n^2 + nm + \sigma_m n}}{(1 - q)(1 - q^2) \cdots (1 - q^n)}.$$

Hence

(3.23) $$F(0) = F_0(0) = \sum_{n=0}^{\infty} \frac{q^{n^2}}{(1-q)(1-q^2)\dots(1-q^n)},$$

and

(3.24) $$F(1) = F_0(1) = \sum_{n=0}^{\infty} \frac{q^{n^2+n}}{(1-q)(1-q^2)\dots(1-q^n)}.$$

Now the true magic occurs. Although a priori $F(0)$ and $F(1)$ would appear to have no relationship to what has arisen before, the celebrated Rogers-Ramanujan identities tell us that

(3.25) $$F(0) = G(q),$$

and

(3.26) $$F(1) = H(q).$$

While the Rogers-Ramanujan identities had been unknown in physics prior to Baxter's pathbreaking work, they had played a significant role in the theory of partitions of integers [1; Ch. 7]. For example, (3.25), the first Rogers-Ramanujan identity, implies that the partitions of n into distinct nonconsecutive summands are equinumerous with the partitions of n into summands whose last digit is 1, 4, 6 or 9. Thus there are five partitions of 9 into distinct nonconsecutive summands (9, 8+1, 7+2, 6+3, 5+3+1) and five into summands whose last digit is 1, 4, 6 or 9 (9, 6+1+1+1, 4+4+1, 4+1+1+1+1+1, 1+1+1+1+1+1+1+1+1).

Substituting (3.25) and (3.26) into (3.16) and (3.17) we find that

(3.27) $$Z = \frac{H(x)G(x^6) - xG(x)H(x^6)}{H(x)}$$

and

(3.28) $$<\sigma_1> = \frac{-xG(x)H(x^6)}{H(x)G(x^6) - xG(x)H(x^6)}.$$

We should note that Baxter is one of only three people (Rogers, Schur, Baxter) who found and proved the Rogers-Ramanujan identities without knowing of them beforehand. Some comments on Baxter's discovery occur in [3].

4. Ramanujan's identities.

We concluded the last section with the identity

(4.1) $$<\sigma_1> = \frac{-G(x)H(x^6)}{H(x)G(x^6) - x\,G(x)H(x^6)}$$

and it might reasonably be asserted that this is magic enough. However, the denominator for $<\sigma_1>$ recalls a one page abstract (Paper 29 in [16]) in which Ramanujan announced that he had found a number of algebraic identities for $G(x)$ and $H(x)$:

(4.2) $$\text{``}H(x)G(x)^{11} - x^2 G(x)H(x)^{11} = 1 + 11x\,G(x)^6 H(x)^6.$$

Another noteworthy formula is

(4.3) $$H(x)G(x^{11}) - x^2 G(x)H(x^{11}) = 1.$$

Each of these formulae is the simplest of a large class."

Turning to Ramanujan's Lost Notebook [17] we find a list of 40 such identities, and, in particular we find (as did Baxter [6; Ch. 14]):

(4.4) $$H(x)G(x^6) - x\,G(x)\,H(x^6) = \frac{P(x)}{P(x^3)},$$

where

(4.5) $$P(x) = \prod_{n=1}^{\infty}(1 - x^{2n-1}),$$

which allows us to obtain Baxter's final formula for $< \sigma_1 >$:

$$< \sigma_1 > = -x\,G(x)H(x^6)P(x^3)/P(x)$$

(4.6) $$= -x \prod_{n=1}^{\infty}(1 - x^n)^{-\lambda_n}$$

where $\lambda = 2$ if $n \equiv \pm 1, \pm 11 \pmod{30}$, $\lambda = 0$ if $n \equiv \pm 2,\ \pm 3,\ \pm 8, \pm 10, 15 \pmod{30}$, $\lambda = 1$ otherwise.

These 40 formulae (exemplified by (4.2), (4.3) and (4.4)) have long seemed tantalizing results, and have attracted the attention of a number of notable mathematicians, among them Rogers [18], Mordell [14], Watson [21], Birch [7], and Bressoud [9].

That these formulae are scarcely to be believed is easily demonstrated if, for example, we translate (4.3) directly into the theory of partitions.

Theorem 4.1 (Bressoud [9]). Let $\pi_1(n)$ denote the number of partitions of n into parts that are either congruent to $\pm 2 \pmod 5$ or $\pm 11 \pmod{55}$. Let $\pi_2(n)$ denotes the number of partitions of n into parts that are either congruent to $\pm 1 \pmod 5$ or $\pm 22 \pmod{55}$. Then $\pi_1(n) = \pi_2(n-2)$ for each $n > 0$.

As an example, take $n = 11$. The five partitions enumerated by $\pi_1(11)$ are 11, 8+3, 7+2+2, 3+3+3+2 and 3+2+2+2+2, while the five enumerated by $\pi_2(9)$ are 9, 6+1+1+1, 4+4+1, 4+1+1+1+1+1, and 1+1+1+1+1+1+1+1+1.

General questions suggested by this Theorem and its possible generalizations are considered in [2]. Bressoud in his thesis [9] was apparently the first to see the partition-theoretic implication of Ramanujan's formulae.

It turns out that even more general results than (4.4) are requried to treat more general models. With this motivation, Baxter discovered a general theta function identity to cover an infinite number of such results. In full generality Baxter's result provides an elegant example of how discoveries in physics lead to theorems with wide number-theoretic application.

5. Baxter's identity.

Baxter's work in physics has often (as we have seen) required use of results for the classical theta functions and the related elliptic functions. An excellent introduction to these techniques is contained in Chapter 15 of his book [6] and can be read independently of the rest of the book.

To illustrate Baxter's achievements in this area we present a general result found by Baxter which has until now only been seen on the forty-fifth page of a long paper devoted to a generalization of the Hard Hexagon Model [4].

The only background we need is Jacobi's triple product identity [1; p. 21]:

(5.1)
$$E(z,q) \equiv \prod_{n=1}^{\infty}(1 - q^{n-1}z)(1 - q^n z^{-1})(1 - q^n)$$
$$= \sum_{n=-\infty}^{\infty} (-1)^n q^{n(n-1)/2} z^n.$$

Theorem 5.1 (Baxter [4; pp. 237–239]).

Let x and y be real numbers such that $|x| < 1$ and

(5.2)
$$y^{m/2} = -\epsilon x^r, \qquad \epsilon = \pm 1$$

where $1 \le m < 2r$ are positive integers. Then for all nonzero z

(5.3)
$$\sideset{}{^*}\sum_{-r<a<r} x^{a(a-1)/2} z^a E(x^a, y) E(\epsilon^{m-1} x^{(2r-m)(r+1)} z^{2r}, x^{2r(2r-m)}).$$
$$= \frac{1}{2}\{E(-z, x)E(z^{-1}, y/x) \pm E(z, x)E(-z^{-1}, y/x)\}$$

The notation $\sum^*$ means that the summation is to be taken over all even a or all odd a; furthermore if the choice is even a then the $+$ sign is chosen on the right and if it is odd a then the $-$ sign is chosen.

PROOF. We begin with

(5.4)
$$J = \sideset{}{^*}\sum_{a=-\infty}^{\infty} \sigma(a)$$

where

(5.5)
$$\sigma(a) = x^{a(a-1)/2} z^a E(x^a, y).$$

Now for any integer u,

$$E(q^{-u}z, q) = \prod_{n=1}^{\infty}(1 - q^{n-u-1}z)(1 - q^{n+u}z^{-1})(1 - q^n)$$

$$= (q - q^{-u}z)\cdots(1 - q^{-1}z)\prod_{n=1}^{\infty}(1 - zq^{n-1})(1 - q^{n+u}z^{-1})(1 - q^n)$$

(5.6)

$$= (-z)^u q^{-u(u_1)/2}\prod_{n=1}^{\infty}(1 - zq^{n-1})(1 - q^n z^{-1})(1 - q^n)$$

$$= (-z)^u q^{-u(u+1)/2}E(z, q).$$

(The above argument assumes that $u \geq 0$; for $u < 0$ we use $E(z, q) = E(q/z, q)$ and repeat the above). From (5.2) and (5.6) it follows directly that

$$\sigma(a + 2kr) = x^{2kra+2k^2r^2-kr}z^{2kr}(-x^a)^{-km}$$

(5.7)

$$\times y^{-km(km-1)/2}\sigma(a)$$

$$= (-\epsilon^{m-1})^k x^{(2r-m)k(a+kr)}z^{2kr}\sigma(a).$$

By the division algorithm we have unique d and k for each α given by

(5.8)
$$a = \alpha + 2kr \quad \text{where} \quad -r \leq a_0 < r$$

(note that α and a have the same parity). We can therefore rewrite (5.4) as

(5.9)
$$J = \sum_{-r \leq \alpha < r}^{*} \sum_{k=-\infty}^{\infty} \sigma(\alpha + 2kr).$$

Applying (5.7) to the inner summand and then invoking (5.1), we find

(5.10)
$$J = \sum_{-r \leq \alpha < r}^{*} \sigma(\alpha)E[\epsilon^{m-1}x^{(2r-m)(r+\alpha)}z^{2r}, x^{2r(2r-m)}].$$

By (5.5) (upon noting that $\sigma(-r) = 0$), we see that J is in fact the left-hand side of (5.3). Thus to conclude the proof we need only identify J with the right-hand side of (5.3).

We now define J_e and J_0 by

(5.11)
$$J_e = \sum_{a=-\infty}^{\infty} \sigma(2a),$$

(5.12)
$$J_0 = \sum_{a=-\infty}^{\infty} \sigma(2a + 1).$$

Consequently

(5.13)
$$J_e + \tau J_0 = \sum_{a=-\infty}^{\infty} \tau^a \sigma(a)$$

where $\tau = \pm 1$ and the summation is over *all* integers a. Hence by (5.5) and (5.1), we see that

(5.14)

$$J_e + \tau J_0 = \sum_{a=-\infty}^{\infty} \tau^a x^{a(a-1)/2} z^a \sum_{k=-\infty}^{\infty} (-1)^k x^{ak} y^{k(k-1)/2}$$

$$= \sum_{k=-\infty}^{\infty} (-\tau)^k z^{-k} (y/x)^{k(k-1)/2} \sum_{j=-\infty}^{\infty} \tau^j z^j x^{j(j-1)/2}$$

$$= E(\tau/z, y/x) E(-\tau z, x),$$

where we have interchanged the a, k summations and set $a = j - k$.

Taking sums and differences of (5.14) with $\tau = 1$ and $\tau = -1$, we can evaluate J_e and J_0 explicitly. Since J is either J_e or J_0, we observe immediately that J is identical with the right-hand side of (5.3). $\square$

The instance of Theorem 5.1 relevant to the Hard Hexagon Model is the case $m = 4$, $\epsilon = -1$, $y = -x^{5/2}$, $z = -x/y = x^{-3/2}$. In this case we group together the terms a and $-a$ and apply the elementary identities

(5.15) $$E(z^{-1}, q) = -z^{-1} E(z, q)$$

(5.16) $$E(z, q) = E(-q z^2, q^4) - z E(-q z^{-2}, q^4).$$

as well as (5.6) to obtain

(5.17)

$$\sideset{}{^*}\sum_{1 \leqslant a \leqslant 4} x^{a(a-4)/2} E(x^a, -x^{5/2}) E(x^{3a}, x^{15})$$

$$= x^{-2} E(-x^{1/2}, x) E(x^{1/2}, x^6).$$

Taking the even case (i.e. $a = 2$ and $a = 4$), we obtain

(5.18)

$$x^{-2} E(x^2, -x^{5/2}) E(x^6, x^{15}) + E(x^4, -x^{5/2}) E(x^9, x^{15})$$

$$= x^{-2} E(-x^{1/2}, x) E(x^{1/2}, x^6).$$

Now in the notation of this section,

(5.19) $$G(q) = \frac{E(q^2, q^5)}{Q(q)},$$

and

(5.20) $$H(q) = \frac{E(q, q^5)}{Q(q)},$$

where

(5.21) $$Q(q) = \prod_{n=1}^{\infty} (1 - q^n).$$

Thus replacing x by q^2, then replacing q by $(-q)$ in (5.18) and then dividing both sides by $Q(q)Q(q^6)$, we find

$$H(q)G(q^6) - q\,G(q)H(q^6)$$
$$= \frac{E(q, q^2)E(-q, q^{12})}{Q(-q)Q(q^6)},$$

which reduces to (4.4) after algebraic simplification of the right-hand side using the infinite product representations of both E and Q.

6. Conclusion.

As we have mentioned before, this has been a limited and sketchy introduction to the relationship between number theory and statistical mechanics. Indeed it appears that this fruitful interaction is gaining in intensity. The interested reader would do well to consult the papers of Date, Jimbo, Kuniba, Miwa and Okado (in particular [10]) where vast machines are developed to solve the most amazing models. Quite likely we will see this work greatly influencing the direction of research in additive number theory.

References

1. G. E. Andrews, *The Theory of Partitions*, Encyclopedia of Math. and Its Applications (Rota, ed.), Vol. 2, Addison-Wesley, Reading, 1976. (Reissued: Cambridge University Press, London and New York, 1985).

2. G. E. Andrews, *Further problems on partitions*, Amer. Math **94** (1987), 437–439.

3. G. E. Andrews and R. J. Baxter, *A motivated proof of the Rogers-Ramanujan identities*, Amer. Math. Monthly **96** (1989), 401–409.

4. G. E. Andrews, R. J. Baxter and P. J. Forrester, *Eight-vertex SOS model and generalized Rogers-Ramanujan-type identities*, J. Stat. Phys. **35** (1984), 193-266.

5. F. C. Auluck and D. S. Kothari, *Statistical mechanics and the partitions of numbers*, Proc. Cambridge Phil. Soc. **42** (1946), 272–277.

6. R. J. Baxter, *Exactly Solved Models in Statistical Mechanics*, Academic Press, London and New York, 1982.

7. B. J. Birch, *A look back at Ramanujan's ntoebooks*, Math. Proc. Cambridge Phil. Soc. **78** (1975), 73–79.

8. N. Bohr and F. Kalckar, *On the transmutation of atomic nuclei by impact of material particles*, I. general theoretical remarks, Kgl. Danske Vid. Selskab. Math. Phys. Medd., 14, No. 10, 1937, 40 pp.

9. D. M. Bressoud, *Proof and generalization of certain identities conjectured by Ramanujan*, Ph. D. thesis, Temple University, 1977.

10. E. Date, M. Jimbo, A. Kuniba, T. Miwa and M. Okado, *Exactly solvable SOS models II: proof of the star-triangle relation and combinatorial identities*, Advanced Studies in Pure Math **16** (1988), 17–122.

11. P. A. MacMahon, *Combinatory Analysis*, Vol. 2, Cambridge Univeristy Press, London, 1916 (Reprinted: Chelsea, New York, 1960.).

12. P. A MacMahon, *Two applications of general theorems in combinatory analysis*, Proc. London Math. Soc (2) **15** (1916), 314–321.

13. H. B. Mann and D. R. Whitney, *On a test of whether one of two random variables is stochastically larger than the other*, Ann. Math. Stat. **18** (1947), 50–60.

14. L. J. Mordell, *Note on certain modular relations considered by Messrs Ramanujan, Darling, and Rogers*, Proc. London Math. Soc. (2) **20** (1922), 408–416.

15. J. J. Sylvester, *A constructive theory of partitions*, Amer. J. Math., 5 (1882), 251–330, 6 (1884) 334–336., (also in Coll. Math. Papers of J. J. Sylvester, Vol. 4, Cambridge University Press, London and New York, 1912, pp. 1–83. [Reprinted: Chelsea, New York, 1974])..

16. S. Ramanujan, *Collected Papers*, Cambridge University Press, London and New York, 1927, (Reprinted: Chelsea, New York, 1962).

17. S. Ramanujan, *The Lost Notebook and Other Unpublished Papers*, Narosa, New Delhi, 1987.

18. L. J. Rogers, *On a type of modular relation*, Proc. London Math. Soc. (2) 19 (1921), 387–397.

19. H. N. V. Temperley, *Statistical mechanics and the partition of numbers, II. the form of crystal surfaces*, Proc. Cambridge Phil. Soc. 48 (1953), 683–697.

20. C. Van Lier and G. E. Uhlenbeck, *On the statistical calculation of the density of the energy levels of the nuclei*, Physica 4 (1937), 531–542.

21. G. N. Watson, *Proof of certain identities in combinatory analysis*, J. Indian Math. Soc. 20 (1933), 57–69.

22. E. M. Wright, *Asymptotic partition formulae, I: plane partitions*, Quart. J. Math. Oxford Ser. 2 (1931), 177–189.

Proceedings of Symposia in Applied Mathematics
Volume **46**, 1992

Number Theory and Dynamical Systems

J. C. LAGARIAS

ABSTRACT. This paper describes the occurrence of number-theoretic problems in dynamical systems. These include Hamiltonian dynamical systems, dissipative dynamical systems, and discrete dynamical systems. Diophantine approximations and continued fractions play an important role.

1. Introduction

The origin of the subject of dynamical systems was the study of trajectories of moving bodies. These motions were specified by differential equations. Problems in celestial mechanics were particularly important, including the occurrence of periodic motions and quasi-periodic motions. In the nineteenth century the subject was called *analytical dynamics* and generally concerned the behavior of the solutions of differential equations having analytic coefficients. In this century the notion of what constitutes a dynamical system has broadened and now includes: differentiable dynamical systems, symbolic dynamics, topological dynamics, discrete dynamical systems, and ergodic theory. *Differentiable dynamical systems* concern differential equations with differentiable (not necessarily analytic) coefficients. Smale (1967) exhibited a general class of differentiable dynamical systems, called Axiom A flows, whose trajectories exhibit chaotic behavior. *Symbolic dynamics* is the representation of trajectories by sequences of symbols. It traces back to work of Hadamard (1898), who used symbols to describe the paths of geodesics on a manifold of constant negative curvature. Artin (1924) used symbolic codings to prove the existence of dense geodesics on such manifolds. Morse and Hedlund (1938) originated the name symbolic dynamics and abstracted a notion of a symbolic dynamical system as a mapping $T: \Sigma \to \Sigma$ on a set Σ of symbol sequences. This in turn led to the development of *topological dynamics* by Hedlund and others, and to the viewpoint that the iteration of a transformation $T: X \to X$ on a compact metric space is a discrete

1980 *Mathematics Subject Classification* (1985 *Revision*). Primary 11K50, 11K60, 58F05, 58F07; Secondary 11J70, 26A18, 54H20, 70F15, 82A60.

This paper is in final form and no version of it will be submitted for publication elsewhere.

© 1992 American Mathematical Society
0160-7634/92 $1.00 + $.25 per page

dynamical system. The use of an associated discrete dynamical system to study the trajectories of a continuous dynamical system is due to Poincaré. Poincaré (1912) also initiated the study of maps of the circle and introduced the concept of rotation number. Finally, *ergodic theory* concerns the behavior of statistical averages of given measurements $f(x(t))$ of a trajectory $x(t)$ as t varies. The notion of ergodicity arose in statistical mechanics, e.g., the ergodic hypothesis of Boltzmann. The mathematical treatment of ergodic theory started with the individual ergodic theorem of Birkhoff (1931).

Number theoretic problems have appeared repeatedly in dynamical systems. This initially seems surprising, since number theory deals with discrete objects. Many connections to number theory arise through the occurrence of periodic and quasi-periodic orbits in dynamical systems. The interaction of motions with two different frequencies depends on Diophantine approximation properties of the frequency ratio ω_1/ω_2. Conversely, one-dimensional Diophantine approximation can be studied as a discrete dynamical system, namely the continued fraction algorithm.

This paper describes four connections between number theory and dynamical systems.

First, the dynamical behavior of continuous systems sometimes leads to the appearance of rational numbers. The Solar System supplies numerous examples.

(1) The axis rotation period of the moon and orbital rotation period for the moon around the earth have ratio 1:1. The axis rotation period and orbital rotation period for Mercury around the sun have ratio 3:2.

(2) The asteroid Pallas has mean orbital period in the ratio 7:18 with that of Jupiter, as observed by Gauss. Gauss published a note in an encrypted form in the bulletin of the Göttingen academy to establish priority. In 1812 he mentioned it in a letter to his friend Bessel, asking him to keep it secret "for the duration." (See Schroeder 1990, p. 171.) In addition, other asteroids have orbital periods commensurable with that of Jupiter in the ratios 1:1 (Trojans) and 3:2 (Hildas).

(3) In 1866 Kirkwood observed that there are very few asteroids having Jupiter period:asteroid period ratios near 3:1, 5:2, and 7:3; these are called *Kirkwood gaps*, see Figure 1.1.

(4) The moons Io, Europa, and Ganymede of Jupiter have mean orbital periods λ_I, λ_E, and λ_G satisfying

$$\lambda_I - 3\lambda_E + 2\lambda_G \doteq 0$$

to within an error of 10^{-9}. (See Peale 1976.)

(5) There is a complicated set of gaps in Saturn's rings; see Figure 1.2. A set of these gaps are related to the moon Mimas by the relation

$$\lambda_{\text{Gap}} - 2\lambda_M + \omega_{\text{Gap}} \doteq 0$$

where ω_{Gap} is the longitude at perihelion of the gap. (See Peale 1976.) Other

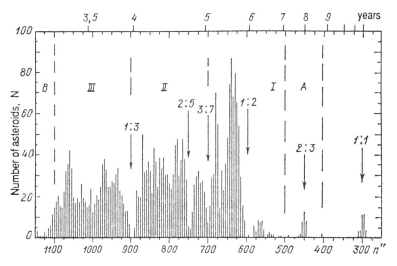

Figure 1.1. Kirkwood gaps. (Reprinted with permission, from V. I. Arnol′d, *Huygens and Barrow, Newton and Hooke, Pioneers in mathematical analysis and catastrophe theory from evolvents to quasicrystals*, Birkhauser, Boston, MA, 1990, p. 78.)

ring gaps are related to other moons, including a recently found 18th moon, named 1981 S13 or Pan, which orbits inside Encke's gap in the outermost ring.

Figure 1.2. Saturn's rings. (Reprinted with permission, from V. I. Arnol′d, *Huygens and Barrow, Newton and Hooke, Pioneers in mathematical analysis and catastrophe theory from evolvents to quasicrystals*, Birkhauser, Boston, MA, 1990, p. 79.)

Some of these commensurabilities are examples of *mode-locking phenomena*; such phenomena are discussed in §6.

Second, number theory arises in stability questions. A major dynamical problem concerns the question of stability of the Solar system: Will the planets continue to orbit in well-defined zones without collision? A mathematical abstraction of this question is: Will point masses representing the planets, evolving according to Newtonian dynamics, remain for all time inside disjoint compact sets (invariant tori), in terms of a coordinate system taking the center of gravity of the Solar System as the origin? Recent numerical simulations suggest that Pluto and its satellite Charon exhibit chaotic motion, with the possibility that their orbital elements may undergo major changes; see Sussman and Wisdom (1988). There is also numerical evidence that the motion of the inner planets in the solar system is chaotic; see Laskar (1990).

This stability problem has been approached mathematically by applying perturbation theory to analyze small perturbations of completely integrable Hamiltonian dynamical systems. Completely integrable systems have regular motion confined within invariant tori. The problem is to assess the effect of perturbations on breaking up such tori. This is the subject matter of KAM (Kolmogorov-Arnol′d-Moser) theory, which says that under small enough perturbations many invariant tori persist. Invariant tori are destroyed because orbits having a rational ratio of periods ("resonances") can inflate small oscillations. Analyzing the effect of "resonances" is the problem of "small divisors," which depends on Diophantine approximation properties of the ratios of two mean orbital periods in the unperturbed system.

Integrable Hamiltonian systems are discussed in §3 and KAM theory and "small divisors" problems in §§4 and 5. The rigorous bounds that have been established, so far, which guarantee the existence of some invariant tori, do not come close to establishing the stability of the major planets in the Solar System. There is a gap between the bounds that are proved and experimental evidence on how far they are valid; this is illustrated for the "standard map" in §4.

Third, it is well known that the ordinary continued fraction algorithm can be treated as a discrete dynamical system on $[0, 1]$. The associated subject is called the metric theory of continued fractions, and is discussed in §2. Here dynamical systems appear in number theory itself. The ordinary continued fraction expansion finds the set of best Diophantine approximations to a real number θ. Another less well-known variant of the continued fraction, called the *additive continued fraction*, finds all the best one-sided Diophantine approximations. The additive continued fraction has an analogous dynamical theory, attached to the *Farey shift map* on $[0, 1]$. Here there is more than one natural invariant measure. In §2 we describe two such measures, namely, an absolutely continuous measure and a singular measure associated to the function $?(x)$ of Minkowski (1904). The additive continued fraction expansion seems more closely connected to many dynamical problems than the ordinary continued fraction; in particular, the mode-locking problems of §6 are related to singular measures.

Fourth, number theoretic structures occur in the study of quasicrystals. Quasicrystals are solids whose X-ray diffraction patterns appear to have sharp peaks at points ("Bragg peaks") and exhibit symmetries forbidden for any periodic lattice in $\mathbb{R}^3$, e.g., 5-fold symmetries are observed. Recent scanning tunneling microscope pictures of pure quasicrystals indicate that physical quasicrystals sometimes have a quasi-periodic structure; these are discussed in §7. Number-theoretic constructions, including *cut-and-project methods* and *substitution sequences*, produce quasi-periodic patterns which have Fourier transforms with such symmetries. They supply possible models for quasicrystalline structures. Substitution rules can also be used directly to construct interesting symbolic dynamical systems, some of which have been extensively studied as mathematical objects.

One of the major open directions relating number theory and dynamical systems concerns finding precise connections between multidimensional dynamical systems and multidimensional Diophantine approximation. It is apparent that the situation in higher dimensions is complicated and known results are fragmentary.

There are many other occurrences of problems arising in dynamical systems that have connections with number theory. In addition various number-theoretic problems have a dynamical flavor, for example, the $3x+1$ problem; see Lagarias (1985). Some general references are given in §0 in the reference list.

2. Continued fractions as dynamical systems

The ordinary continued fraction expansion $\theta_0 = [a_0, a_1, \ldots, a_n, \ldots]$ of a real number has *partial quotients* a_n and *partial remainders* θ_n computed by the recursion

$$(2.1) \qquad \theta_n = a_n + 1/\theta_{n+1}$$

where $a_n = \lfloor \theta_n \rfloor$ is the largest integer not exceeding θ_n. The process halts if some $\theta_n = a_n$, which occurs exactly when θ is rational. The *convergents* $p_n/q_n = [a_0, a_1, \ldots, a_n]$ are given by $p_{-1}/q_{-1} = 1/0$, $p_0/q_0 = a_0/1$, and

$$(2.2) \qquad \frac{p_{n+1}}{q_{n+1}} = \frac{a_n p_n + p_{n-1}}{a_n q_n + q_{n-1}}.$$

They are exactly the set of *best two-sided Diophantine approximations* to θ, i.e., p/q is a convergent if and only if $0 \le |q\theta - p| < |q'\theta - p'|$ holds whenever $0 < q' < q$. The partial remainders θ_n satisfy $\theta_{n+1} = T(\theta_n)$ where $T: [0, 1] \to [0, 1]$ is the *Gauss map*

$$(2.3) \qquad T(x) = \{1/x\} \qquad (\text{mod } 1),$$

where $\{y\} = y - \lfloor y \rfloor$ is the fractional part of y. By convention $T(0) = 0$. (See Figure 2.1.)

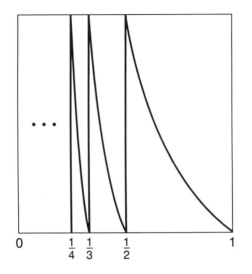

Figure 2.1. Gauss map

Gauss already was aware of the fact that $d\mu_G = dt/(1+t)$ is an invariant measure for T, i.e.,

$$\int_{[a,b]} \frac{dt}{1+t} = \int_{T^{-1}[a,b])} \frac{dt}{1+t}$$

for all intervals $[a, b]$.

Using ergodic theory terminology, the continued fraction algorithm is a *skew-product* $(T, A, d\mu_G)$ over the *base space* $(T, 1/\log 2 \cdot dt/(1+t))$ together with a *fiber map* $A: [0, 1] \to \mathbb{Z}^+$ describing the partial quotients, with

$$A(\theta) = k \quad \text{if } 1/(k+1) < \theta \leq 1/k$$

and $A(0) = 0$.

A. Khinchin (1935), (1936) and P. Levy (1936) developed a metric theory of continued fractions, which treats θ as a random variable on $[0, 1]$ drawn with the Gauss measure $d\mu_G = 1/\log 2 \cdot dt/(1+t)$. The Gauss map with the Gauss measure is ergodic, and one has

THEOREM 2.1 (Khinchin). *If θ is drawn from $[0, 1]$ with the Gauss measure $1/\log 2 \cdot dt/(1+t)$ then the partial quotients of $\theta = [0, a_1, a_2, \ldots]$ are identically distributed with*

(2.4) $\text{Prob}[a_k = m] = \log_2(1 + 1/m(m+2))$.

The successive a_n's are not independent, but the correlation between a_n and a_{n+j} decays rapidly as j increases. This can be used to show that the Gauss map is strongly mixing for the measure $d\mu_G$, i.e., that

$$\lim_{j \to \infty} \mu_G(\mathsf{E} \cap T^{-j}\mathsf{F}) = \mu_G(\mathsf{E})\mu_G(\mathsf{F})$$

for all measurable subsets E, F of $[0, 1]$; see Philipp (1967). One also can deduce such facts as

THEOREM 2.2 (Khinchin-Levy Theorem). *There is a set of Lebesgue measure one in* $[0, 1]$ *such that all* θ *in this set have*

$$(2.5) \qquad \lim_{n \to \infty} \left(\prod_{k=1}^{n} a_k \right)^{1/n} = \prod_{m=1}^{\infty} \left(1 + \frac{1}{m(m+2)} \right)^{\log m / \log 2} \doteq 2.6855$$

and

$$(2.6) \qquad \lim_{n \to \infty} \frac{1}{n} \log(q_n) = \frac{\pi^2}{12 \log 2} \doteq 1.1866.$$

The Gauss measure is the unique absolutely continuous invariant measure for the Gauss map. In particular, if one iterates any other absolutely continuous invariant measure repeatedly by the Gauss map, it will converge to the Gauss measure—a fact that Gauss asserted in a letter to Laplace. This was proven by Kuzmin in 1928. In fact this convergence is exponentially fast.

THEOREM 2.3 (Kuzmin-Levy Theorem). *Let* $d\lambda_n$ *be the* nth *iterate of Lebesgue measure* λ *under the Gauss map* T. *There exists* $\delta < 1$ *and* c_0 *such that*

$$1 - c_0 \delta^n \le \lambda_n(\mathbf{J})/\mu_G(\mathbf{J}) \le 1 + c_0 \delta^n$$

for any open interval $\mathbf{J}$ *in* $[0, 1]$.

P. Levy showed that $\delta < .68$ and Wirsing (1974) determined the best constant $\delta = 0.3035663\dots$. It is interesting that this result can be proved by a procedure analogous to "renormalization" in statistical mechanics; see Mayer (1990).

The *additive continued fraction algorithm* is a variant of the continued fraction algorithm that finds all the intermediate convergents to θ as well as the ordinary continued fraction convergents. The *intermediate convergents* of the ordinary continued fraction algorithm are fractions $(lp_n + p_{n-1})/(lq_n + q_{n-1})$ having $1 \le l < a_n$. These are exactly the *best one-sided Diophantine approximations* to θ, i.e., those fractions p/q such that if $p/q < \theta$ then $0 < q\theta - p < q'\theta - p'$ whenever $1 \le q' < q$ and $p'/q' < \theta$, while if $p/q > \theta$ then $q'\theta - p' < q\theta - p < 0$ whenever $1 \le q' < q$ and $p'/q' > \theta$.

The additive continued fraction has a simple description in terms of the Farey tree pictured in Figure 2.2. The Farey tree $\mathscr{F}$ is an infinite binary tree whose vertices are labeled by all rationals p/q in the open interval $(0, 1)$ as follows. Two fractions p_1/q_1 and p_2/q_2 are *Farey neighbors* if $p_1 q_2 - p_2 q_1 = \pm 1$, i.e., if the interval $[p_1/q_1, p_2/q_2]$ has length $1/q_1 q_2$. For two Farey neighbors, their *mediant* is

$$(2.7) \qquad \frac{p_1}{q_1} \oplus \frac{p_2}{q_2} = \frac{p_1 + p_2}{q_1 + q_2}.$$

This fraction is always in lowest terms and is called their *Farey sum*. The Farey tree starts with the unit interval $[\frac{0}{1}, \frac{1}{1}]$ having mediant $\frac{1}{2} = \frac{0}{1} \oplus \frac{1}{1}$, and the mediant $\frac{1}{2}$ is the label of the root node of the binary tree. Now split

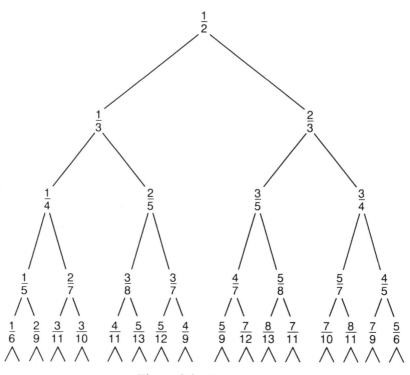

Figure 2.2. Farey tree

the unit interval at its mediant to obtain two subintervals $[\frac{0}{1}, \frac{1}{2}]$ and $[\frac{1}{2}, \frac{1}{1}]$, which have mediants $\frac{1}{3}$ and $\frac{2}{3}$. These mediants label the leaves at level one of the Farey tree. Continuing to split at mediants, at level n we obtain a partition of $[0, 1]$ in 2^n subintervals whose mediants label the Farey tree at level n. We have

LEMMA 2.4. *Every rational* $p/q \in (0, 1)$ *can be expressed uniquely as a Farey sum* $p/q = p_1/q_1 \oplus p_2/q_2$. *Every such rational occurs uniquely as a vertex of the Farey tree.*

To each vertex p/q of the Farey tree is associated the interval $[p_1/q_1, p_2/q_2]$. The *additive continued fraction convergents* of θ are exactly the sequence of vertices of the Farey tree encountered in taking the sequence of nested intervals that contain θ at each level of the Farey tree. This sequence is unique if θ is irrational; there are exactly two such sequences if θ is rational.

There is a symbolic dynamics attached to the Farey tree, in which each edge of the tree is labeled by L (for Left) or R (for Right) according to whether it gives the left or right subinterval of the interval being split. The sequence of edges that one follows from the root in following a set of nested intervals for θ is called a *Farey symbol sequence* for θ. If one sets

(2.8)
$$L = \begin{bmatrix} 1 & 0 \\ 1 & 1 \end{bmatrix}, \qquad R = \begin{bmatrix} 1 & 1 \\ 0 & 1 \end{bmatrix},$$

then the additive continued fraction convergents can be encoded using matrix multiplication.

There is also a metric theory for the additive continued fraction algorithm, which is associated to the *Farey shift map* $U: [0, 1] \rightarrow [0, 1]$ given by

(2.9)
$$U(x) = \begin{cases} x/(1-x) & \text{if } 0 \le x \le \frac{1}{2}, \\ (1-x)/x & \text{if } \frac{1}{2} \le x \le 1. \end{cases}$$

See Figure 2.3. There is also a symbolic dynamics associated to iterates of U, which keeps track of whether a point falls in $[0, \frac{1}{2})$ (labeled with symbol A) or in $[\frac{1}{2}, 1]$ (labeled B).

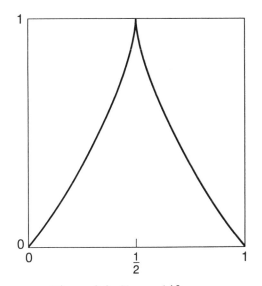

Figure 2.3. Farey shift map

THEOREM 2.5. *The symbolic dynamics of the Farey shift map for $\theta \in (0, 1]$ essentially encodes the additive continued fraction expansion of θ, in the following sense: There is a finite automaton that converts an $\{A, B\}$ symbol sequence for θ to an $\{L, R\}$ symbol sequence and vice-versa.*

The Farey shift map actually encodes θ using the matrices

(2.10)
$$A = \begin{bmatrix} 1 & 0 \\ 1 & 1 \end{bmatrix}, \qquad B = \begin{bmatrix} 0 & 1 \\ 1 & 1 \end{bmatrix}.$$

The Farey shift map has the absolutely continuous invariant density dt/t, and it is ergodic with respect to this density; see Parry (1962). This density has infinite mass, which implies that the metric entropy of the mapping

$(U, dt/t)$ is zero. (Metric entropy has a complicated definition; see Petersen 1983.)

The Farey shift, however, has another invariant measure $d?$, which is the "derivative" of Minkowski's ?-function. The ?-function of Minkowski (1904) is pictured in Figure 2.4.

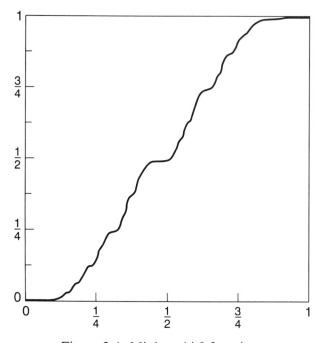

Figure 2.4. Minkowski ?-function.

Minkowski introduced this function as a monotonic function on $[0, 1]$, which gives a one-to-one order-preserving map from all real quadratic irrationals in $[0, 1]$ into the set of rationals with denominator not a power of 2, e.g., $?((\sqrt{5} - 1)/2) = 2/3$. The simplest definition of $?(\theta)$ is that if $\theta = [0, a_1, a_2, a_3, \ldots]$ then

(2.11) $$?(\theta) = \sum_{k=1}^{\infty} (-1)^{k+1} 2^{-(a_1 + \cdots + a_k - 1)}.$$

Salem (1943) showed that $?(x)$ is a singular function, so that $d?$ is a singular measure. Kinney (1960) showed that $d?$ is concentrated on a set of Hausdorff dimension α, with

$$\alpha = \frac{1}{2} \left(\int_0^1 \log_2(1 + x) d? \right)^{-1},$$

in the sense that, for any set S of Hausdorff dimension less than α one has $\int_S d? = 0$, while there is a set S of Hausdorff dimension α with $\int_S d? = 1$.

Computer calculations have determined that

(2.12) $$0.8746 < \alpha < 0.8749.$$

The measure $d?$ is characterized by the property that the dynamical system $(U, d?)$ has maximal entropy. It is known that the metric entropy of a continuous map of the interval never exceeds its *topological entropy*

(2.13) $$h_{\text{top}}(U) = \lim_{n \to \infty} \frac{1}{n} \log(\text{Fix}(U^{(n)}))$$

where $\text{Fix}(F)$ counts the number of fixed points of a map F. Clearly $\text{Fix}(U^{(n)}) = 2^n$ so that

$$h_{\text{top}}(U) = \log 2.$$

The Minkowski measure $d?$ turns out to assign equal weight to every edge of the Farey tree, and Theorem 2.5 then implies that $(U, d?)$ is just the Bernoulli shift on the two letters $\{R, L\}$; hence $(U, d?)$ has metric entropy $\log 2$. This has the following consequence.

THEOREM 2.6. *If θ is drawn from $[0, 1]$ according to the singular measure $d?(x)$, then the partial quotients of $\theta = [0, a_1, a_2, \ldots]$ are independent and identically distributed with*

$$\text{Prob}[a_k = m] = 2^{-m}.$$

The Farey tree construction arises in renormalization schemes to analyze mode-locking phenomena, described in §6, and also in analyzing rotation orbits of circle maps $f: S^1 \to S^1$ of degree $d > 1$; see Goldberg and Tresser (1991).

There are several other connections between continued fractions and dynamical systems. First, a real number θ can be described by a sequence of symbols describing the elements of the lattice $\mathbb{Z}^2$ which are adjacent to the ray $\{(t, \theta t): t \in \mathbb{R}^+\}$ in $\mathbb{R}^2$. This viewpoint traces back to H. J. S. Smith in the 1870s. It is described in Series (1985a) and in McIlroy's paper, since it is useful in computer graphics. This description also arises as a special case of the cut-and-project method of constructing one-dimensional quasicrystals described in §7. Second, there is a well-known use of continued fractions to code geodesics on the modular surface $\mathbb{H}/\text{SL}(2, \mathbb{Z})$ and other surfaces of constant negative curvature; see Series (1982), (1985b). Third, Arnoux and Nogueira (1991) indicate how to reformulate some continued fraction transformations as one-to-one area-preserving maps, which are analogous to completely integrable Hamiltonian systems.

A nice description of the additive continued fraction appears in Richards (1981). Properties of the Farey shift are surveyed in Lagarias and Peres (1992).

3. Integrable Hamiltonian dynamical systems

The dynamical systems of classical mechanics can be put in Hamiltonian form. The Hamiltonian formulation has two important features: it directly identifies conserved quantities of the flow, and it is the classical starting point to obtain quantum mechanical systems; see Dirac (1950). The second-order differential equations of Newtonian dynamics can be converted to a Hamiltonian system of first-order differential equations. (Note however that Hamilton actually developed this formalism in connection with his work in geometrical optics, rather than mechanics.)

The *Hamiltonian function* $H(\mathbf{p}, \mathbf{q}, t)$ describes the dynamics, where $t \in \mathbb{R}$ is a time variable and $(\mathbf{p}, \mathbf{q})$ lie on a $2n$-dimensional real manifold $\mathcal{M}$, locally identifiable with $\mathbb{R}^{2n}$, called "phase space." Here $\mathbf{p} \in \mathbb{R}^n$ are traditionally called *position variables* and $\mathbf{q}$ *momentum variables*. The equations of motion are

$$(3.1) \qquad \frac{d\mathbf{p}}{dt} = -\frac{\partial H}{\partial \mathbf{q}}, \qquad \frac{d\mathbf{q}}{dt} = \frac{\partial H}{\partial \mathbf{p}}.$$

One consequence is

$$\frac{dH}{dt} = \frac{\partial H}{\partial t} + \left\langle \frac{\partial H}{\partial \mathbf{p}}, \frac{d\mathbf{p}}{dt} \right\rangle + \left\langle \frac{\partial H}{\partial \mathbf{q}}, \frac{d\mathbf{q}}{dt} \right\rangle = \frac{\partial H}{\partial t}.$$

In particular if $H = H(\mathbf{p}, \mathbf{q})$ is autonomous (independent of t-variable), i.e., the time variable does not appear explicitly, then $dH/dt = 0$, so that H is a *conserved quantity* under the flow (3.1).

In what follows we suppose that $H(\mathbf{p}, \mathbf{q})$ is autonomous. The simplest example is

$$H(\mathbf{p}, \mathbf{q}) = \sum_{i=1}^{n} \frac{1}{2}p_n^2 + V(\mathbf{q}),$$

where the first term represents kinetic energy and the second term potential energy; this Hamiltonian prescribes conservation of energy.

The phase space $\mathcal{M}$ is endowed with a distinguished two-form (*symplectic form*), which locally has the form

$$(3.2) \qquad \omega = \Sigma \, dp_i \wedge dq_i$$

and is closed ($d\omega = 0$). Any transformation of coordinates $\Phi(\mathbf{p}, \mathbf{q}) = (\mathbf{p}', \mathbf{q}')$ that preserves this form is called a *canonical transformation*; it preserves the form (3.1) of the differential equations in $(\mathbf{p}', \mathbf{q}')$ coordinates. A key feature of Hamiltonian flows Φ_t defined by $\Phi(\mathbf{p}(0), \mathbf{q}(0))) = (\mathbf{p}(t), \mathbf{q}(t))$ is that they preserve area, as given by the symplectic form ω. This implies *Liouville's theorem*: Hamiltonian flows preserve volume in phase space. In consequence, one may consider area-preserving maps as discrete analogues of Hamiltonian systems.

The symplectic form induces a Lie algebra structure on smooth functions $K(\mathbf{p}, \mathbf{q})$ on the manifold, via the operation of *Poisson bracket*, given by

$$(3.3) \qquad \{K_1, K_2\} = \sum_{i=1}^{n} \left(\frac{\partial K_1}{\partial \mathbf{p}_i} \frac{\partial K_2}{\partial \mathbf{q}_i} - \frac{\partial K_2}{\partial \mathbf{p}_i} \frac{\partial K_1}{\partial \mathbf{q}_i} \right).$$

A function K is said to *commute with* H or be *in involution with* H if

(3.4) $$\{K, H\} \equiv 0.$$

In that case $K(\mathbf{p}, \mathbf{q})$ is also a conserved quantity of the system (3.1); such K are also sometimes called *integrals of the motion*.

The number of *degrees of freedom* for a time-independent Hamiltonian system $H(\mathbf{p}, \mathbf{q})$ is n, and that of a time-dependent Hamiltonian system $H(\mathbf{pq}, t)$, which is periodic in t,

$$H(\mathbf{p}, \mathbf{q}, t + 1) \equiv H(\mathbf{p}, \mathbf{q}, t),$$

is by convention $n + \frac{1}{2}$.

A Hamiltonian system is said to be *globally completely integrable* if there exist n independent commuting integrals of the motion

(3.5) $$\{K_i, H\} = 0, \qquad 1 \le i, j \le n,$$

with nonvanishing Jacobian everywhere in position space. (Usually one of the K_i is H.) What is the effect of this? A solution curve of (3.1) satisfies n independent conservation laws

$$K_i(\mathbf{p}, \mathbf{q}) = 0, \qquad 1 \le i \le n,$$

which are all transverse and, hence, is constrained to an n-dimensional submanifold. In fact, under a suitable canonical transformation it can be taken to be the $\mathbf{q}$-variables, with $H(\mathbf{p}, \mathbf{q}) = H(\mathbf{p})$, and Hamilton's equations become

$$\frac{\partial \mathbf{p}}{\partial t} = 0, \qquad \frac{\partial \mathbf{q}}{\partial t} = \frac{\partial H}{\partial \mathbf{p}}.$$

Now a further canonical coordinate change gives, locally,

(3.6) $$\frac{\partial p_i}{\partial t} = 0, \qquad \frac{\partial q_i}{\partial t} = \omega_i p_i$$

for constants $\{\omega_i : 1 \le i \le n\}$. In the form (3.6) the variables p_i are called *action variables* and the variables q_i *angle* variables.

It is often a difficult problem to find action-angle coordinates for a completely integrable system.

One consequence of (3.6) is that the flow lines of a completely integrable system fit together smoothly. This contrasts with nonintegrable systems, which can exhibit "chaotic" motion; see §4.

THEOREM 3.1 (Arnol'd). *A globally completely integrable Hamiltonian system has phase space manifold diffeomorphic to* $\mathbb{R}^n \times \Sigma$ *with* $\Sigma \cong \mathbb{R}^{n-k} \times \mathbb{T}^k$, *for some* $0 \le k \le n$.

The most studied case of this is $\Sigma \cong \mathbb{T}^n$; the variables $\mathbf{q}$ are then all periodic, which justifies the term *angle variables*. See Arnol'd, Kozlov, and Neistadt (1988), p. 110 for references and a proof sketch.

Theorem 3.1 can be rephrased as asserting that the phase space is diffeomorphic to $\mathbb{R}^n / \Lambda$, where Λ is a discrete subgroup of $\mathbb{R}^n$, i.e., Λ is a lattice and k is the rank of Λ. The connection of completely integrable systems with number theory apparently arises from the discrete subgroup Λ. In the case of the torus $\mathbb{T}^n = \mathbb{R}^n/\mathbb{Z}^n$, the flow curves can sometimes be analyzed in terms of Diophantine approximation.

There is also an extension of Theorem 3.1 to *locally completely integrable systems*, which are systems for which there is a set of n independent commuting integrals $\{K_i : 1 \le i \le n\}$ that may have vanishing Jacobian at some points of position space. A point $\mathbf{x}$ is *nondegenerate* if the Jacobian of the set $\{K_i : 1 \le i \le n\}$ does not vanish at $\mathbf{x}$. The local version is: *The closure of the orbit of an integrable system passing through a nondegenerate point is diffeomorphic to* $\mathbb{R}^{n-k} \times \mathbb{T}^k$ *for some* $0 \le k \le n$. Here k depends on the point. For example, the frictionless pendulum is locally completely integrable, and its orbits in phase space are $\mathbb{T}$ or $\mathbb{R}$ according to whether the pendulum swings back and forth or goes "over the top." Ito (1991) gives canonical forms for integrable Hamiltonian systems in a neighborhood of a singular point.

All Hamiltonian systems with one degree of freedom are locally completely integrable. For two or more degrees of freedom integrable Hamiltonian systems are very rare; generic Hamiltonian systems are not completely integrable (see Arnol'd, et al. (1988)). Completely integrable systems appear to play some special role in quantization, in the theory of "geometric quantization."

We describe four examples of completely integrable systems.

EXAMPLE 1 (Simple Harmonic Oscillator). The Hamiltonian is

$$H = \sum_{i=1}^{n} \tfrac{1}{2}p_i^2 + \sum_{i=1}^{n} \tfrac{1}{2}w_i q_i^2 ,$$

where the first term represents kinetic energy and the second potential energy. A set of Hamiltonians in involution is

$$F_i = \tfrac{1}{2}p_i^2 + \tfrac{1}{2}w_i q_i^2 , \qquad 1 \le i \le n.$$

This Hamiltonian describes a set of n noninteracting particles. For each particle separately one has

$$\frac{dp_i}{dt} = -w_i q_i , \qquad \frac{dq_i}{dt} = p_i.$$

which is simple harmonic motion, so the phase space is $\mathbb{T}^n$.

More generally, since all 1-dimensional Hamiltonian systems are locally completely integrable, except at isolated points, any system

$$H = \sum_{i=1}^{n} H_i(p_i , q_i)$$

is locally completely integrable.

EXAMPLE 2 (Two-body problem). This system in $\mathbb{R}^3$ has $n = 6$ degrees of freedom; these are accounted for by conservation of energy, of momentum, and of angular momentum $(1 + 3 + 2)$.

The relevance of this integrable system in celestial mechanics is that some N-body problems can be viewed as perturbations of a 2-body problem if two masses are very large relative to the others, for example, the sun and Jupiter. In the Solar system example (4) in the introduction, the variable ω_{Gap} appears in terms of action-angle coordinates (3.6) for the two-body problem (Saturn, Mimas).

EXAMPLE 3 (Geodesic flow on a Riemannian manifold). The simplest case is geodesic flow on the torus $\mathbb{T}^n = \mathbb{R}^n/\mathbb{Z}^n$. Here $\mathbb{T}^n$ is a flat manifold with the induced metric from $\mathbb{R}^n$. In describing the behavior of geodesics one encounters the problem of multidimensional Diophantine approximations, where the initial conditions encode the numbers to be approximated. Geometrically the number of $\mathbb{Z}$-independence conditions for $(\alpha_1, \ldots, \alpha_n)$ determines the dimension of the torus which is the closure of the geodesic.

EXAMPLE 4 (Toda Lattice). The Toda lattice is an infinite chain of particles having exponential nearest neighbor interactions given by

$$\ddot{q}_n = e^{q_{n+1}-q_n} - e^{q_n-q_{n-1}}.$$

The Hamiltonian function is

$$H = \sum_n \tfrac{1}{2}p_n^2 + \sum_n (e^{q_n-q_{n-1}} - 1).$$

It is an infinite-dimensional system but has finite-dimensional analogues obtained by either imposing periodic boundary conditions or by fixing end points; see Moser (1975). Flaschka (1974) observed that one of the finite-dimensional analogues can be put into the Lax-pair form

$$\dot{\mathbf{L}}(t) = [\mathbf{B}(t), \mathbf{L}(t)],$$

where $\mathbf{L}(t)$ is a symmetric tridiagonal matrix, and

$$\mathbf{L}(t) = \begin{bmatrix} a_1(t) & b_1(t) & & & 0 \\ b_1(t) & a_2(t) & \ddots & & 0 \\ & \ddots & & \ddots & \\ & & \ddots & & b_{n-1}(t) \\ 0 & 0 & & b_{n-1}(t) & a_n(t) \end{bmatrix},$$

and

$$\mathbf{B}(t) = \begin{bmatrix} 0 & -b_1(t) & & & 0 \\ b_1(t) & 0 & \ddots & & 0 \\ & \ddots & & \ddots & \\ & & \ddots & & -b_{n-1}(t) \\ 0 & 0 & & b_{n-1}(t) & 0 \end{bmatrix},$$

where

$$a_k(t) = \frac{1}{2} e^{(q_k - q_{k-1})/2}$$

$$b_k(t) = -\frac{1}{2} p_k = -\frac{1}{2} \frac{d}{dt}(q_k).$$

The flow $\mathbf{L}(t)$ is isospectral, and the coefficients of the characteristic polynomial of $\mathbf{L}(t)$ form a full set of commuting Hamiltonians for the flow, making it completely integrable. Symes (1982) observed a close connection between this form of the Toda flow and the QR-algorithm for numerically finding the eigenvalues of a symmetric matrix. This example has phase space $\cong \mathbb{R}^n$.

Further examples of completely integrable systems are given by certain flows on the Jacobian varieties of algebraic curves, which naturally have the structure of (complex) tori. The solutions of such differential equations can often be written in terms of elliptic functions. There are also integrable systems associated to certain interior-point methods for linear programming; see Bayer and Lagarias (1989).

As a final topic, we observe that one may relax the usual notion of completely integrable system by not requiring that all conserved quantities commute with each other. Define a *G-completely integrable system* to be a set $\{K_1, \ldots, K_m\}$ of functions on a symplectic manifold $\mathcal{M}$ with $K_1 = K$, which form an m-dimensional Lie algebra under Poisson bracket and whose Jacobian is of full-rank at each point of $\mathcal{M}$. It appears that an analogue of Theorem 3.1 holds where the closure of each orbit has structure similar to a simply connected Lie group modulo the action of a discrete subgroup; see Arnol'd, Koslov, and Neishtadt (1988) p. 110. It seems probable that the occurrence of a discrete subgroup for such systems will eventually lead to developments related to number theory.

4. KAM theory and small divisors

KAM theory concerns the behavior of the orbits of a "nearly integrable" dynamical system. A "nearly integrable" system is one with Hamiltonian

$$(4.1) \qquad H_\varepsilon = H_0(\mathbf{p}, \mathbf{q}) + \varepsilon H_1(\mathbf{p}, \mathbf{q}, \varepsilon),$$

where H_0 is completely integrable and ε is a small parameter. We will suppose that all orbits of H_0 have the form $\mathbb{T}^n$. The problem is approached by perturbation theory in the parameter ε. Under what conditions does an orbit for $\varepsilon = 0$ smoothly deform into an orbit for $\varepsilon = \varepsilon_0$? Such problems are directly motivated by celestial mechanics, in studying the question of the stability of the solar system.

We assume that action-angle variables $(\mathbf{I}, \boldsymbol{\theta})$ are given, where $\mathbf{I} = (I_1, \ldots, I_n) \subset B \subseteq \mathbb{R}^n$ with B compact and $\boldsymbol{\theta} = (\theta_1, \ldots, \theta_n) \pmod{2\pi} \in \mathbb{T}^n$, so that the Hamiltonian $H_0 = H_0(\mathbf{I})$ depends only on the action variables. Hamilton's equations are

$$(4.2a) \qquad \frac{d\mathbf{I}}{dt} = \mathbf{0}, \qquad \frac{d\boldsymbol{\theta}}{dt} = \frac{\partial H_0}{\partial \mathbf{I}}.$$

The perturbation $\varepsilon H_1(\mathbf{I}, \boldsymbol{\theta}, \varepsilon)$ yields

(4.2b)
$$\frac{d\mathbf{I}}{dt} = -\varepsilon \frac{\partial H_1}{\partial \boldsymbol{\theta}}, \qquad \frac{d\boldsymbol{\theta}}{dt} = \frac{\partial H_0}{\partial \mathbf{I}} + \varepsilon \frac{\partial H_1}{\partial \mathbf{I}}.$$

In the perturbed system the quantities $(I_1, \ldots, I_n)$ are no longer exactly conserved, and one wishes to know how they change over time, i.e., do they remain bounded? Here the rotational quantities $\boldsymbol{\theta}$ are viewed as *fast variables* and the quantities $\mathbf{I}$ as *slow variables*, which may gradually evolve over time. The *frequencies* of the unperturbed system are

(4.3)
$$\omega_i = \frac{\partial H_0}{\partial I_i}\Big|_{\mathbf{I}}.$$

A system is *nonresonant* if $(\omega_1, \ldots, \omega_n)$ are linearly independent over $\mathbb{Z}$.

A much-used approach is the method of *averaging*, which approximates the system (4.2b) averaged over the fast variables.

The detailed approach to the averaging method is to construct suitable changes of coordinates that formally eliminate, to some degree of accuracy, the fast variables. Such coordinate changes started with Lindstedt, with contributions by Delaunay, Poincaré, and many others. One seeks $(\mathbf{I}, \boldsymbol{\theta}) \to (\mathbf{J}, \boldsymbol{\psi})$ expressed as a formal series

(4.4)
$$\mathbf{I} = \mathbf{J} + \varepsilon \mathbf{u}_1(\mathbf{J}, \boldsymbol{\psi}) + \varepsilon^2 \mathbf{u}_2(\mathbf{J}, \boldsymbol{\psi}) + \cdots,$$
$$\boldsymbol{\theta} = \boldsymbol{\psi} + \varepsilon \mathbf{v}_1(\mathbf{J}, \boldsymbol{\psi}) + \varepsilon^2 \mathbf{v}_2(\mathbf{J}, \boldsymbol{\psi}) + \cdots,$$

where the functions $\mathbf{u}_i$ and $\mathbf{v}_i$ are periodic $(\text{mod } 2\pi)$ in $\boldsymbol{\psi}$. In the new coordinate system one wants independence of $\boldsymbol{\psi}$ variables:

(4.5)
$$\frac{d\mathbf{J}}{dt} = \varepsilon \mathbf{F}_0(\mathbf{J}) + \varepsilon^2 \mathbf{F}_1(\mathbf{J}) + \cdots,$$
$$\frac{d\boldsymbol{\psi}}{dt} = \boldsymbol{\omega}(\mathbf{J}) + \varepsilon \mathbf{G}_0(\mathbf{J}) + \cdots.$$

Given a 2π-periodic function in the $\boldsymbol{\psi}$-variables $u(\mathbf{J}, \boldsymbol{\psi})$, expanding it in Fourier series gives

(4.6)
$$u(\mathbf{J}, \boldsymbol{\psi}) = f_0(\mathbf{J}) + \sum_{\substack{\mathbf{k} \in \mathbb{Z}^n \\ \mathbf{k} \neq \mathbf{0}}} h_k(\mathbf{J}) \exp(i\langle \mathbf{k}, \boldsymbol{\psi} \rangle).$$

Assuming that H_1 is a sufficiently smooth function, one can expand it in a power series in ε, then expand each term in Fourier series (4.6), and recursively solve (4.5) to determine explicit expressions for $\mathbf{u}_i$, $\mathbf{v}_i$ in (4.4) up to order r, say. In the resulting expressions the quantities $\langle \mathbf{k}, \boldsymbol{\omega} \rangle^{-1}$ appear, where

$$\langle \mathbf{k}, \boldsymbol{\omega} \rangle = k_1 \omega_1 + \cdots + k_n \omega_n,$$

where $\mathbf{k} \in \mathbb{Z}^n$ and each $|k_i| \leq r$. Values $\langle \mathbf{k}, \boldsymbol{\omega} \rangle$ close to 0 are called "small denominators" and may make the formal series (4.4) diverge. Lindstedt's series usually diverges.

Kolmogorov (1954) and Arnol'd (1963) obtain a modified recursion to get transformations (4.4), which exhibits a 'quadratic convergence' in that after r steps all terms ε^k with $k < 2^r$ are eliminated. The influence of the small denominators can be controlled if ω satisfies a bound, of the form

$$|\langle \mathbf{k}, \boldsymbol{\omega} \rangle| > c\|\mathbf{k}\|^{-\nu},$$

which holds for all nonzero $\mathbf{k} \in \mathbb{Z}^n$ and $\nu > n-1$. We say that a Hamiltonian system H_0 is *nondegenerate* at energies $\mathbf{I}$ if $\det(\partial^2 H_0 / \partial \mathbf{I}^2) \neq 0$.

THEOREM 4.1 (Kolmogorov's theorem). *If in* (4.2) *the unperturbed system* $H_0(\mathbf{I})$ *is nondegenerate, then for a sufficiently small Hamiltonian perturbation* $\varepsilon H_1(\mathbf{I}, \boldsymbol{\theta}, \varepsilon)$ *most nonresonant invariant tori do not vanish but are slightly deformed, so that the phase space of the perturbed system contains invariant tori, which contain dense trajectories that have* n *independent frequencies of motion. The measure of the complement of the set of invariant tori approaches zero at least as fast as* $\sqrt{\varepsilon}$ *as* $\varepsilon \to 0$.

Kolmogorov stated this result when H_0 and H_1 are analytic functions, but it has been extended to apply when H_1 is only C^r with $r > 2n$ by Pöschel. The $\sqrt{\varepsilon}$ result on the complementary set is due to later authors, see Arnol'd, et al. (1988), Chapter 5. In particular, on these invariant tori the action variables $\mathbf{I}$ remain bounded for all time.

KAM-type phenomena can also be studied in the simpler discrete setting of area-preserving maps. A much-studied example is the *standard map* or *Chirikov-Taylor map* $S(r, \theta) = (r', \theta')$ on $\mathbb{T}^2$ given by

$$\begin{aligned}
r' &= r - \frac{K}{2\pi} \sin(2\pi\theta) \quad (\text{mod } 1) \\
\theta' &= \theta + r' \quad (\text{mod } 1).
\end{aligned}$$
(4.7)

Here K is a coupling parameter. For $0 \leq K < 1$ this is a *twist map* since $\partial \theta'/\partial r > 0$; see Mather (1982). For $K = 0$ the closure of the orbits of S of rotation number θ consists of *invariant circles* for all irrational θ in $[0, 1]$; see Figure 4.1. (*Rotation number* is defined by equation (6.3); the irrationality of θ is a nonresonance condition.) As K increases invariant circles are destroyed. At the instant of destruction the closure of an orbit becomes a perfect set, like a Cantor set; such sets are called *cantori*. For larger values of K there exist orbits whose closure has positive measure.

The problem of how long an invariant circle of a given rotation number θ exists as K varies has been extensively treated. An analogue of KAM theory has been applied to establish that at least one invariant circle exists for $0 \leq K < 1/30$. The most difficult rotation number to destroy appears to be $\theta = (\sqrt{5} - 1)/2$, apparently since it is the hardest number in $[0, 1]$ to approximate by rationals.

A heuristic method to estimate when circles break up was developed by Greene (1979). The idea is that invariant circles are broken by a sequence

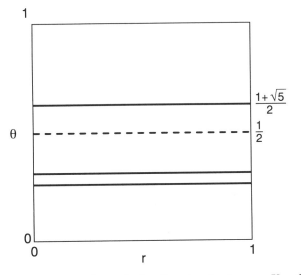

Figure 4.1. Invariant circles for standard map $K = 0$.

of stable resonances nearby. The linear stability of a periodic orbit of a map S of period q is measured by its *residue*

$$(4.8) \qquad R = \tfrac{1}{4}(1 - \mathrm{Tr}(\nabla(S^{(q)}))).$$

The Poincaré-Birkhoff twist theorem says that, for each rational p/q, an area-preserving twist map has at least two periodic orbits of rotation number p/q—one with residue greater than zero, one with residue less than zero. (They are critical points of an associated action.) Generically there are two, having residues $R_{p/q}^{-} \leq 0 \leq R_{p/q}^{+}$, say. Greene observed empirically that for orbits of period q_n, the nth ordinary continued fraction convergent to θ, if, as $n \to \infty$

   (i) $R_{p_n/q_n}^{\pm} \to 0$ then a smooth invariant circle exists;

   (ii) $R_{p_n/q_n}^{\pm} \to \pm\infty$ then no invariant circle exists.

For $\theta = (\sqrt{5} - 1)/2$, computer experiments using this heuristic indicate that the critical value K_{∞} at which the invariant circle of rotation number θ disappears is $K_{\infty} = 0.971635406$; see Figure 4.2.

Above this value of K there are apparently no invariant circles.

5. Iteration of analytic functions and small divisors

In addition to KAM theory, small divisor problems arise in studying normal forms for the iteration of analytic function around a fixed point. This situation is analogous to a Hamiltonian system with one degree of freedom. Consider an analytic function $F(z)$ having a fixed point at 0, so that

$$(5.1) \qquad F(z) = a_1 z + a_2 z^2 + a_3 z^3 + \cdots.$$

What is the behavior of $F(z)$ under iteration in a neighborhood of $z = 0$? If $|a_1| < 1$ then, for small z, $|F(z)| < |z|$ and $F^{(k)}(z_0) \to 0$ as $k \to \infty$

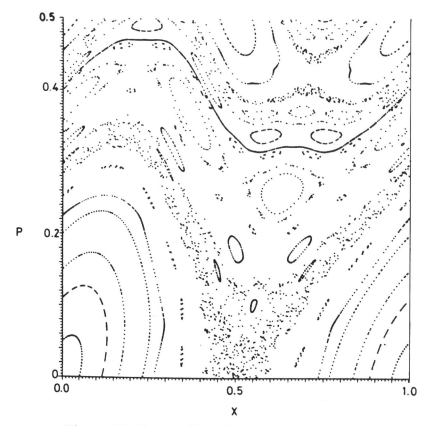

P

X

Figure 4.2. Some orbits of the standard map at $K =$ 0.971635406. The golden circle orbit starts at 0.4 at left. (Reprinted from R. S. MacKay, J. D. Meiss, and I. C. Percival, *Transport in Hamiltonian systems*, Phys. D **13** (1984), 67.)

and 0 is an *attracting fixed point*. If $|a_1| > 1$ then $|F(z)| > |z|$ for small z and 0 is a *repelling fixed point*. The interesting case is $|a_1| = 1$, which is called an *indifferent fixed point*. Here the higher order terms in (2.1) become relevant. Let $a_1 = e^{2\pi i\theta}$. The simplest case is the *pure rotation*.

$$(5.2) \qquad\qquad F_\theta(z) = e^{\pi i\theta} z.$$

The behavior of F_θ under iteration is different according as θ is rational or irrational. In both cases circles $C_r = \{z : |z| = r\}$ are invariant sets; however, if $\theta = p/q$ is rational then $F^{(q)}(z) \equiv z$ and every point lies in a finite orbit, while if θ is irrational, then the orbit of every point $z \in C_r$ is dense in C_r, in fact, uniformly distributed (Weyl). What happens for general

$$(5.3) \qquad\qquad F(z) = e^{\pi i\theta} z + a_2 z^2 + a_3 z^3 + \cdots?$$

For some such F an analytic change of variable

$$(5.4) \qquad \varphi(z) = z + b_2 z^2 + b_3 z^3 + \cdots$$

conjugates $F(z)$ *to a rotation*, i.e.,

$$(5.5) \qquad \varphi^{-1}(F(\varphi(z))) = e^{\pi i \theta} z.$$

When this occurs, the qualitative behavior of iterates of $F(z)$ is the same as the rotation $F_\theta(z)$; there are invariant circles, which are analytic curves, in a neighborhood of 0.

Equation (5.5) is called the *Schröder functional equation* after Schröder (1871). Schröder came to this problem from studying general iterative schemes to find solutions of equations; cf. Schröder (1870). Newton's method is a special case of such an iterative scheme. This is interesting since Newton's method proves important in studying the Schröder functional equation itself.

In the case of rational $\theta = p/q$ the functional equation (5.5) cannot necessarily be solved, even formally, while for any irrational θ it always has a formal power series solution (5.4) where the coefficients b_i are found recursively:

$$(5.6) \qquad \begin{aligned} b_2 &= (e^{2\pi i\theta} - e^{\pi i\theta})^{-1} a_2 \\ b_3 &= (e^{3\pi i\theta} - e^{\pi i\theta})^{-1}(a_3 + 3(1 - e^{\pi i\theta})a_2 b_2 + 2(e^{3\pi i\theta} - e^{2\pi i\theta})b_2^2) \end{aligned}$$

The resulting power series, however, may have radius of convergence zero. The problem in determining the radius of convergence arises from the "small denominators" $(e^{n\pi i\theta} - e^{\pi i\theta})^{-1}$ appearing in the formal coefficients b_n. When $\varphi(z)$ has a positive radius of convergence one says that $F(z)$ is *analytically conjugate to a rotation*. For certain values of θ every convergent series (5.3) is analytically conjugate to a rotation, while for other values of θ this is not so. Siegel (1942) showed that the *Diophantine condition*: There is a constant $\beta > 0$, such that

$$(5.7) \qquad |\theta - p/q| \geq 1/q^\beta \quad \text{for all } (p, q) \in \mathbb{Z}^2, \, q \neq 0,$$

guarantees that all convergent series (5.3) are analytically conjugate to $e^{2\pi i\theta} z$. Siegel also gave examples of an irrational θ where there exists some $F(z)$ not analytically conjugate to a rotation.

Brjuno (1971) gave a wider sufficient condition on θ that guarantees analytic conjugacy.

THEOREM 5.1 (Brjuno). *Let θ have the ordinary continued fraction expansion $[0, a_1, a_2, a_3, \ldots]$ with convergents p_n/q_n. All functions $F(z) = e^{2\pi i\theta} z + a_2 z^2 + \cdots$ analytic in a neighborhood of 0 are analytically conjugate to the rotation $e^{\pi i\theta} z$, provided that θ satisfies*

$$(5.8) \qquad \sum_{i=1}^{\infty} q_n^{-1} \log q_{n+1} < \infty.$$

For such θ Brjuno showed that a sequence of successive polynomial approximations $\varphi_n(z)$ converge uniformly to a solution $\varphi(z)$ to (5.5) near 0, thus guaranteeing analyticity of $\varphi(z)$. The $\varphi_n(z)$ are constructed by a Newton-method-like process.

Recently Yoccoz (1988) showed that Brjuno's condition is also necessary, which completely solves this problem.

How do the iterates of a point near 0 behave for $F(z)$ when it is not analytically conjugate to a rotation? Baker and Rippon (1984) show that there exists an irrational θ for which the entire function

$$F(z) = e^{2\pi i \theta}(e^z - 1)$$

has a point z_0 whose iterates are *dense* in $\mathbb{C}$. Such z_0 can clearly be found in any neighborhood of 0, and invariant circles near 0 do not exist.

Further information can be found in Martinet (1981) and Herman (1987).

6. Mode-locking

The *mode-locking phenomenon* for dissipative systems is that weakly coupled oscillators tend to synchronize their motion, i.e., their modes of oscillation acquire $\mathbb{Z}$-linear dependencies. This arises when there are two or more periodic motions with competing frequencies with some sort of a dissipative coupling force between them. The motion of the three moons of Jupiter (given in §1) is an example of mode-locking. Mode-locking occurs also in some simple Josephson junction circuits; see Bak, Bohr, and Høgh-Jensen (1985).

Typical motions of such dissipative systems may include a mixture of quasi-periodic motion, mode-locked motion, and chaotic motion, depending on the initial frequencies and the degree of nonlinearity.

An extensively studied simple model is the *sine circle map*

(6.1) $$x_{n+1} = f_{K,\Omega}(x_n),$$

where

(6.2) $$f_{K,\Omega}(x) = x + \Omega - \frac{K}{2\pi}\sin(2\pi x),$$

where $K \geq 0$ is a parameter; this map was introduced in Arnol'd (1965). Here x_n (mod 1) may be thought of as an angle, measuring motion around the circle $\mathbb{T} = \mathbb{R}/\mathbb{Z}$ (with units chosen so that a full angle is 1 rather than 2π.) The map (6.1) is a lift of it to the universal cover $\mathbb{R}$. For $K = 0$, the map is

$$x_{n+1} \equiv x_n + \Omega \quad (\text{mod } 1),$$

which is a pure rotation. The two competing frequencies of the motion are the *bare rotation rate* Ω and the rotation rate 1 associated to the driving term $\sin 2\pi x$, and K is the coupling constant. For $0 \leq K < 1$ the function $f_{K,\Omega}: \mathbb{R} \to \mathbb{R}$ is monotone increasing, with positive derivative everywhere.

The *critical value* is $K = 1$, where $f_{1,\Omega}$ is still monotone increasing but has inflection points at all $x \in \mathbb{Z}$.

For $0 \leq K \leq 1$ the map $f_{K,\Omega}$ has a well-defined *rotation number*

(6.3) $$r(\Omega, K) = \lim_{n \to \infty} (x_n - x_0)/n,$$

which measures the average speed of a point around the circle $\mathbb{T}$. This limit exists and is independent of the starting point x_0 (but depends on Ω and K), by a result of Herman (1977), because $f_{K,\Omega}$ is a monotone twist map. The mode-locking phenomenon here is that for $0 < K \leq 1$ the rotation number $r(\Omega, K)$ takes rational values on intervals, while a given irrational value is taken at a point. (See Figure 6.1.) A rational rotation $r(\Omega, K) = p/q$ is a $\mathbb{Z}$-linear dependency between the rotation number r and the driving frequency 1.

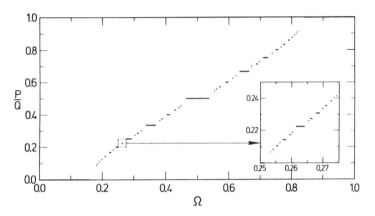

Figure 6.1. Devil's staircase—rotation numbers for sine circle map at $K = 1$. Enlarged box shows self-similar structure. (Reprinted with permission, from M. Hogh-Jensen, P. Bak and T. Bohr, *Complete devil's staircase, fractal dimension, and universality of mode-locking structure in the circle map*, Phys. Rev. Lett. **50** (1983), 1638.)

As the nonlinearity parameter K increases, the intervals become larger, and, as indicated in Figure 6.2, the regions (Ω, K), having a given rational rotation number p/q assume the shape of tongues, which are sometimes called "Arnol'd tongues."

Mode-locking occurs because the monotone increasing nature of the function (6.2) means that whenever f_K has a periodic point $(\mathrm{mod}\, 1)$, i.e., a solution to

(6.4) $$f_{K,\Omega}^{(p)}(x) = x + q,$$

by (6.3) it must have $r(\Omega, K) = p/q$. One sees by plotting $f_{K,\Omega}^{(p)}(x) \ (\mathrm{mod}\, 1)$ versus $x \ (\mathrm{mod}\, 1)$ that since the graph changes continuously (on the torus)

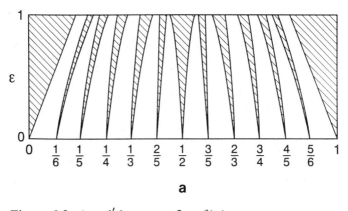

Figure 6.2. Arnol'd tongues for $f(x) = x + a + \varepsilon \cos \pi x$ (mod 1).

as Ω moves, there is an interval of values of Ω where a solution x to (6.4) exists.

For $0 < K < 1$ there is thus a mixture of possible motions consisting of mode-locked motion for Ω in the locked intervals, and quasi-periodic motion at an irrational rotation rate $r(\Omega, K)$ otherwise. What happens at $K = 1$? Høgh-Jensen, Bak, and Bohr (1983) and also Ostlund, et al. (1983) studied the resulting devil's staircase structure of the rotation rates $r(\Omega, 1)$ as Ω increases; see Figure 6.1. They observed an apparent self-similar structure in this staircase and conjectured that the mode-locked intervals have Lebesgue measure one and that the set of Ω such that $r(\Omega, 1)$ is irrational has Hausdorff dimension strictly between zero and one. They numerically estimated this Hausdorff dimension to be 0.8700 ± 0.0002 by a "renormalization" method. Swiatek (1988) rigorously proved the first part of their conjectures.

THEOREM 6.1 (Swiatek). *For the sine circle map with* $K = 1$,

$$x_{n+1} = x_n + \Omega + \frac{1}{2\pi} \sin(2\pi x_n),$$

where the set of Ω, *having rational rotation number* $r(\Omega, 1)$, *has Lebesgue measure one.*

Swiatek's proof uses projective invariants (cross ratios) to keep track of scaling.

Several authors have introduced "renormalization schemes" to estimate universal parameters associated to the critical value $K = 1$, such as the Hausdorff dimension of the unlocked set; cf. Cvitanovic, et al. (1985), Ostlund and Kim (1985), Procaccia, et al. (1987), Feigenbaum (1988), Sinai and Khanin (1988), and Kim and Ostlund (1989). These schemes look at the behavior of the locked intervals at the nth level of the Farey tree $\mathscr{F}_n$ and examine asymptotics of scaled quantities as $n \to \infty$. The Farey shift map

U of §2 appears explicitly in Feigenbaum (1988). One can consider these schemes as assigning a measure to the nodes of $\mathscr{F}_n$, which leads to a limiting singular measure μ_∞ on $[0, 1]$, which encodes "universal" information. It appears likely that this measure is an invariant measure for the Farey shift map. Some recent work on such universal behavior appears in Sullivan (1989) and Khanin (1991).

What happens for $K > 1$? The maps $f_{K,\Omega}$ are no longer monotone and iterates of (5.1) can exhibit chaotic motion in the following sense. Define

$$(6.5a) \qquad r_+(\Omega, K) = \sup_{x_0} \left\{ \limsup_{n\to\infty}(x_n - x_0)/n \right\},$$

$$(6.5b) \qquad r_-(\Omega, K) = \inf_{x_0} \left\{ \liminf_{n\to\infty}(x_n - x_0)/n \right\}.$$

There exist values of K with $r_+(\Omega, K) > r_-(\Omega, K)$, i.e., maps have *rotation intervals* rather than rotation numbers. There are still self-similar structures for $K > 1$; see Jakobson (1988). Furthermore, Brucks and Tresser (1991) have shown for fixed $K > 1$ that if as Ω varies no $f_{K,\Omega}$ has frequency locking at a rational rotation number p/q (i.e., p/q always lies in a rotation interval of nonzero width), then the same property holds for all rationals p'/q' lying above p/q in the Farey tree.

The width of Arnol'd tongues of rotation number p/q near $K = 0$ decreases like K^q; see Arnol'd (1983) and Jonker (1990). This is related to "forbidden zones" in solid state physics.

An important open problem concerns what happens for mode-locking in systems with three or more competing frequencies. Is multidimensional Diophantine approximation relevant to analysis of this question? An initial discussion appears in Kim and Ostlund (1985), (1986).

A general discussion of mode-locking appears in Schroeder (1990), p. 171 ff.

7. Quasicrystals and substitution sequences

Shechtman, et al. (1984) startled the physics world when they found an alloy of aluminum and manganese whose X-ray diffraction patterns in certain directions exhibited discrete sets of sharp spots characteristic of a crystal, in which some of the patterns exhibited five-fold symmetry; see Figure 7.1. The sharp spots indicated long-range order in the atomic structure of the alloy, and the observed symmetries suggested icosahedral symmetry in this structure. It had long been believed, however, that any crystalline arrangement must be periodic in three independent directions, and such arrangements are classified by 230 space groups, none of which possesses any five-fold symmetries; see Milnor (1976). This apparent contradiction led to a rush of activity to understand the structure of such materials, which are now called *quasicrystals*.

The apparent contradiction can be resolved by dropping the assumption of periodicity in three directions. Some kind of quasi-periodic long-range order must exist to produce sharp spots in the X-ray diffraction pattern, which justifies the name quasicrystals. But what sort of quasi-periodic patterns would lead to such diffraction patterns?

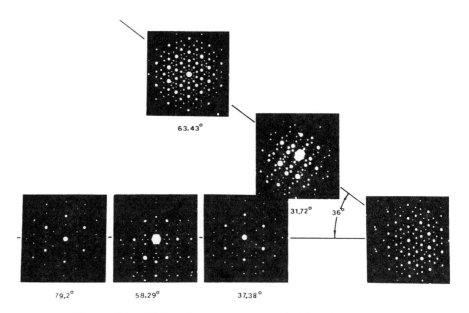

Figure 7.1. Diffraction patterns exhibiting icosahedral symmetry. (Reprinted with permission, from D. Shechtman, I. Blech, D. Gratias and J. W. Cahn, *Metallic phase with long-range orientational order and no translational symmetry*, Phys. Rev. Lett. **53** (1984), 1952.)

There already existed an extensive mathematical theory of nonperiodic tilings in the plane, exemplified by the well-known Penrose tilings, which are analyzed in de Bruijn (1981). This was used to produce possible tiling models for quasicrystals, e.g., Katz and Duneau (1986), Levine and Steinhardt (1986), Socolar and Steinhardt (1986), and many others. For a mathematical viewpoint see Cahn and Taylor (1987) and Senechal and Taylor (1990).

Despite the attractiveness of the quasi-periodic model, there was debate for several years as to whether quasicrystals could somehow be explained without dropping the assumption of periodicity in three directions; however, recent scanning tunneling microscope pictures of two-dimensional quasicrystals $Al_{65}Co_{20}Cu_{15}$, having 10-fold symmetry, by Kortan, et al. (1990), exhibit absence of periodicity; see Figure 7.2. (This material is regarded as a two-dimensional quasicrystal because it is periodic in the third direction.) These pictures are consistent with a Penrose tiling-type quasicrystal model.

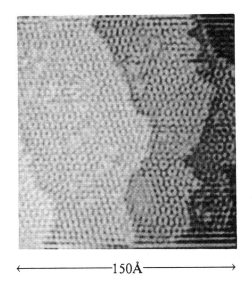

$\longleftarrow\!\!\!\!-\!\!\!\!-\!\!\!\!-\!\!\!\!150\text{Å}\!\!\!\!-\!\!\!\!-\!\!\!\!\longrightarrow$

Figure 7.2. Scanning tunneling microscope image of $AL_{65}Co_{20}Cu_{15}$ quasicrystal, in 2-D decagonal phase (enhanced picture). (Reprinted with permission, from A. Kortan, R. S. Becker, F. A. Thiel and H. S. Chen, *Real-space atomic structure of a two-dimensional decagonal quasicrystal*, Phys. Rev. Lett. **64** (1990), 201.)

The discovery of quasicrystals led to a review of the notion of what a crystal is, and a reexamination of many incommensurate structures occurring in crystals; see Janssen and Janner (1987). The International Union of Crystallography has now redefined the word 'crystal' to mean a solid having an essentially discrete X-ray diffraction pattern.

There are two major questions to answer concerning quasicrystals. The first concerns their structure, i.e., the arrangement of atoms in space. The second concerns their existence, i.e., how an aperiodic arrangement of atoms can occur as a locally energy-minimizing state. We describe mathematical results related to the first question. Concerning the second question, at present there is no satisfactory theory explaining the existence of ordinary crystals, let alone quasicrystals; see Simon (1984), problem 11 and Radin (1991). Some work related to the existence question includes Penrose (1989) and Radin (1986).

Now we formulate a mathematical problem, following Cahn and Taylor (1987). We wish to model the X-ray diffraction pattern of sharp spots. An X-ray diffraction pattern is essentially a Fourier transform that determines magnitudes of Fourier coefficients $|g(\boldsymbol{\xi})|^2$ but loses all phase information. The presence of sharp spots means that the Fourier transform can be modeled as a sum of delta-functions taken with various weights. The function $f(\mathbf{x})$, whose Fourier transform

$$(7.1) \qquad g(\boldsymbol{\xi}) = \int_{\mathbb{R}^3} f(\mathbf{x}) \exp(i\langle \mathbf{x}, \boldsymbol{\xi} \rangle)\, d\mathbf{x}$$

is being taken, is the set of atoms in the solid, which is modeled as a set of delta functions with constant weights

$$(7.2) \qquad f(\mathbf{x}) = \sum_i \delta(\mathbf{x}_i).$$

Also, since we have a closely packed solid, we expect that there are constants c_0 and c_1 such that for each atom $\mathbf{x}_i$ one has

$$(7.3) \qquad c_0 \le \min_{j \ne i} \|\mathbf{x}_i - \mathbf{x}_j\| \le c_1,$$

i.e., the packing of atoms is fairly even. Now one can ask: Which arrangements $\{\mathbf{x}_i\}$ satisfying (7.3) have a Fourier transform that appears to be a discrete set of spots? If the $\{\mathbf{x}_i\}$ are arranged in a lattice Λ, then the Fourier transform will have δ-functions at the points $\{\boldsymbol{\xi}_j\}$ of its dual lattice Λ^* and will be discrete. It appears plausible that if the $\{\mathbf{x}_i\}$ are not arranged in a finite number of cosets of a lattice Λ, then the transform $g(\boldsymbol{\xi})$, even if it consists of δ-functions, cannot be at a discrete set of points $\{\boldsymbol{\xi}_j\}$, i.e., the set $\{\boldsymbol{\xi}_j\}$ must have limit points. X-ray diffraction patterns like those in Figure 7.1, however, are still possible in such cases because the set having large values of $|g(\boldsymbol{\xi}_j)|^2$ can be discrete, and the other diffraction pattern points are too faint to be seen in the pattern.

Thus we have arrived at a harmonic analysis problem of studying all functions of a given type whose Fourier transforms are of a given type. To make the problem precise we must specify a class of allowable functions (i.e., distributions), which includes various functions of type (7.2) and which is closed under the Fourier transform. The space of tempered distributions is one suitable space in which to study cut-and-project methods, as indicated in Porter (1988). Another convenient function space is that of de Bruijn (1973), which is identical with the Gelfand-Shilov space $S(\frac{1}{2}, \frac{1}{2})$ according to van Eijndhoven (1987). This space was used by de Bruijn (1986) to study Fourier transforms of quasicrystal distributions (7.2).

We shall adopt here the following mathematical definition of a quasicrystal in $\mathbb{R}^n$. It is a sum of delta functions (7.2) at a discrete set of points satisfying (7.3), whose Fourier transform is a weighted sum of δ-functions

$$\hat{f}(\boldsymbol{\xi}) = \sum_i c(\boldsymbol{\xi}_i)\delta(\boldsymbol{\xi}_i),$$

having the property that for any $\varepsilon > 0$ the set

$$S(\varepsilon) = \{\boldsymbol{\xi}_i : |c(\boldsymbol{\xi}_i)| > \varepsilon\}$$

has all points separated by a distance $r(\varepsilon)$ depending on ε. This definition includes crystals as a subset of quasicrystals. Note that there is no single agreed-on definition of quasicrystal; see Senechal and Taylor (1990).

Now we simplify further by considering the one-dimensional case, i.e., we allow quasi-periodic structure in only one direction. Thus we have a row of atoms on the line $\mathbb{R}$. We also simplify (7.3) further by allowing only a finite set of possible distances $D_1, D_2, \ldots, D_k$ between adjacent atoms on the line. The possible sets $\{x_i\}$ allowed are tilings of the line $\mathbb{R}$ with tiles drawn from a finite set of tile lengths.

This one-dimensional problem has some special features not present in higher dimensions. It had been studied long before quasicrystals were discovered, under the name "modulated crystals." In fact one-dimensional lattice models with aperiodic potentials (here the potential plays the role of tile distance) have been used to model various problems in solid-state physics, in particular, the discrete Frankel-Kontorova model (see Aubry and Le Daeron 1983), the almost-Mathieu equation, the ANNNI model, and the Schrödinger equation with almost periodic potential (see Bellissard, Lima, and Testard 1985 and Simon 1982).

Several classes of one-dimensional aperiodic tilings have some number-theoretic structure, namely those constructed by the cut-and-project method and those constructed by substitution rules. These classes overlap but neither includes the other. The cut-and-project method is more general in that it gives rise to an uncountable number of different tilings, while there are only a countable number of substitution rule tilings.

The cut-and-project method is to project points from a higher-dimensional lattice Λ in $\mathbb{R}^n$ onto a line l in $\mathbb{R}^n$, where one only projects points that are within a given distance D of the line in $\mathbb{R}^n$, or more generally, in a specified cylinder around the line that has an arbitrarily shaped compact convex cross-section. In the simplest case the lattice is $\mathbb{Z}^2$, the line passes through 0, and the tiling is determined by D and the slope of the line l. If l passes through $\mathbf{v} = (1, \theta)$ and $D = 1$, the properties of the Fourier transform are intimately associated to the continued fraction expansion of θ, as in Series (1985a). The cut-and-project method easily extends to higher dimensions. For example, deBruijn (1981) showed how to construct Penrose tilings by projecting a subset of a certain five-dimensional lattice onto a plane. By picking such lattices Λ to have a given symmetry group, one can arrange for the projected set of points to have a Fourier transform exhibiting that symmetry group. For the study of Fourier transforms of cut-and-project sets, see de Bruijn (1986). The concept of cut-and-project sets appears already in an abstract form in Meyer (1972), where they are called *models*.

Now we turn to another class of one-dimensional tilings, those described by substitution rules; this is illustrated by the *Fibonacci tiling*. Here one has two tile lengths, labeled A and B, and the substitution rules

(7.4)
$$\begin{cases} A \to AB \\ B \to A. \end{cases}$$

Starting with 1, successive substitutions produce

$$A\ B$$
$$A\ B\ A$$
$$A\ B\ A\ A\ B$$
$$A\ B\ A\ A\ B\ A\ B\ A$$
$$A\ B\ A\ A\ B\ A\ B\ A\ A\ B\ A\ A\ B$$
$$\cdots$$

If we let $\mathbf{W}_n$ denote the nth string produced in this process, then the number of symbols in $\mathbf{W}_n$ is the nth Fibonacci number F_n, as easily follows from the observation that

$$\mathbf{W}_n = \mathbf{W}_{n-1}\mathbf{W}_{n-2} \quad \text{(concatenation)}.$$

The process of producing $\mathbf{W}_n$ from $\mathbf{W}_{n-1}$ is called *inflation*; one might think of it as expanding $\mathbf{W}_{n-1}$ and examining it at a level of detail (using a microscope) where the substitution rules tell how a word looks when "inflated." This process of inflation is specified by an integer matrix called a transition matrix, whose rows count the number of symbols of each type produced by the substitution rules. For the Fibonacci tiling rules (7.4) the transition matrix is $[\begin{smallmatrix} 1 & 1 \\ 1 & 0 \end{smallmatrix}]$. The largest eigenvalue of the transition matrix, which is $(1 + \sqrt{5})/2$, specifies the exponential growth rate of the number of symbols in the words $\mathbf{W}_n$ as $n \to \infty$.

Now let $\mathbf{W}_\infty$ denote the infinite limit word constructed from the $\mathbf{W}_n$. It is a *fixed point* of the substitution rules (7.4), i.e., it is self-similar under inflation. This self-similarity property, together with the irrationality of the transition matrix eigenvalues, guarantees the aperiodicity of the resulting tiling of the half line. One may tile all of $\mathbb{R}$ by adding a reflection of this tiling to itself.

In fact the Fibonacci tiling can also be obtained by the cut-and-project method from the line l in $\mathbb{R}^2$ passing through 0 and having slope $\theta = (\sqrt{5} + 1)/2$ by orthogonally projecting all lattice points at distance at most θ^{-1} above or θ^{-2} below l. There are two tiles, a long tile and a short tile, which get the labels A and B, respectively; see Figure 7.3.

A second example of a substitution rule tiling is the Thue-Morse tiling, obtained from the constant-length substitution rules:

(7.5)
$$\begin{cases} A \to BA \\ B \to AB. \end{cases}$$

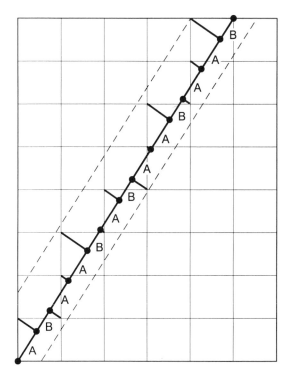

Figure 7.3. Fibonacci quasicrystal.

This yields, starting with A

$$A\ B$$
$$A\ B\ B\ A$$
$$A\ B\ B\ A\ B\ A\ A\ B$$

$$\cdots$$

The resulting fixed point

$$\mathbf{W}_\infty^* = A\ B\ B\ A\ B\ A\ A\ B\ B \cdots$$

is called the Thue-Morse sequence, after Thue (1912). It is aperiodic, although this substitution rule has the transition matrix $[\begin{smallmatrix} 1 & 1 \\ 1 & 1 \end{smallmatrix}]$, which has integer eigenvalues 2 and 0. This sequence has the characteristic property that it contains no blocks of symbols of the form $\mathbf{WWW}$, where $\mathbf{W}$ is any finite string of zeros and ones.

The *paper folding sequences* studied by Mendes-France and van der Poorten (1981) are also closely related to substitution sequences.

For substitution rules we have the problem of determining which such rules yield tilings that have a Fourier transform consisting of delta-functions (discrete spectrum), as opposed to singular continuous spectrum or mixed spectrum, which can also occur. It turns out that the Fibonacci tiling gives a quasicrystal, so it is sometimes called the *Fibonacci quasicrystal*, but the Thue-Morse sequence has both a discrete spectrum and some singular continuous spectrum and does not give a quasicrystal according to the definition above. Bombieri and Taylor (1987) showed that if a substitution rule has a transition matrix whose characteristic polynomial is irreducible and whose largest root is a Pisot number, then it has a discrete spectrum and can also be obtained by a cut-and-project construction. A *Pisot number* is a real algebraic integer α with $\alpha > 1$ such that all its algebraic conjugates are inside the open unit disk $|z| < 1$. Pisot numbers have special Diophantine approximation properties, for example, the set $\alpha^n \pmod 1$ has only 0 as a limit point. They play an interesting role in harmonic analysis; see Salem (1963) and Meyer (1972).

Substitution rules and their fixed points have been studied at length for over fifty years in a variety of mathematical contexts. One can construct dynamical systems from them, which have been studied in their own right; see Queffélec (1987).

One can also study more complicated block-substitution rules. We close with the following unsolved problem, which appears in Keane (1990).

KEANE'S SELF-GENERATING SEQUENCE PROBLEM. Consider the infinite sequence of ones and twos

$$\mathbf{W}^* = \underline{22}\,\underline{11}\,\underline{2}\,\underline{1}\,\underline{22}\,\underline{1}\,\underline{22}\,\underline{11}\,\underline{2}\,\underline{11}\,\underline{22}\,\underline{1}\,\underline{2}\,\underline{11}\,\underline{2}\,\underline{11}\,\underline{2}\,\underline{1}\cdots,$$

which has the self-generating property that the lengths of its blocks of consecutive symbols generate the same sequence. Prove or disprove that the sequence $\mathbf{W}^*$ has a limiting frequency of occurrence of 1's which is $\frac{1}{2}$; that is, if s_n counts the number of 1's in the first n symbols of $\mathbf{W}^*$, then

$$(7.6) \qquad\qquad\qquad \lim_{n \to \infty} (s_n)/n = \tfrac{1}{2}.$$

Computations convincingly support the truth of (7.6). Note that $\mathbf{W}^*$ is uniquely determined by its first digit 2; if one started with 1 the sequence $1\mathbf{W}^*$ results.

The sequence $\mathbf{W}^*$ was first proposed by Kolakowski (1966). Dekking (1981) showed that $\mathbf{W}^*$ cannot be generated by single-letter substitution rules; however, $\mathbf{W}^*$ can be generated from the initial word 22 using the block-substitution rules:

$$\begin{cases} 22 \to 2211 \\ 21 \to 221 \\ 12 \to 211 \\ 11 \to 21 \end{cases}.$$

Keane's problem looks simple, but the problem of determining if fixed points of general block substitution sequences have limiting digit frequencies is known to be undecidable.

Acknowledgments

I am grateful for helpful comments and references from P. Baldwin, A. R. Calderbank, M. Keane, and C. Skinner. M. Senechal carefully read §7 and made a number of valuable suggestions.

§0. GENERAL REFERENCES

1. V. I. Arnol′d, V. V. Kozlov, and A. I. Neistadt (1988), *Mathematical aspects of classical and celestial mechanics*, Dynamical Systems III (V. Arnol′d, ed.), Springer-Verlag, New York.
2. T. Bedford, M. Keane, and C. Series, eds. (1991), *Ergodic theory, symbolic dynamics and hyperbolic spaces*, Oxford Univ. Press, Oxford.
3. M. Dodson and J. M. Vickers, eds. (1989), *Number theory and dynamical systems*, Cambridge Univ. Press, Cambridge.
4. H. Furstenberg (1981), *Recurrence in ergodic theory and combinatorial number theory*, Princeton Univ. Press, Princeton, NJ.
5. M. Jaric, Ed. (1989), *Introduction to the mathematics of quasicrystals*, Academic Press, New York.
6. J.-M. Luck, P. Moussa and M. Waldschmidt, eds. (1990), *Number theory and physics*, Proc. les Houches, France 1989, Springer-Verlag, New York.
7. M. Schroeder (1990), *Fractals, chaos, power laws: Minutes from an infinite paradise*, Freeman, New York.

§1. INTRODUCTION

1. V. I. Arnol′d (1990), *Huygens and Barrow, Newton and Hooke*, Birkhäuser, Boston, MA.
2. E. Artin (1924), *Eine mechanisches system quasiergodischen Bahnen*, Abh. Math. Sem. Univ. Hamburg, vol. 3, pp. 170–175.
3. G. D. Birkhoff (1927), *Dynamical systems*, Amer. Math. Soc. Colloq. Publ. vol. 9, Amer. Math. Soc., Providence, RI, pp. 359–379.
4. G. D. Birkhoff (1931), *Proof of the ergodic theorem*, Proc. Nat. Acad. Sci. U.S.A. **17**, 656–660.
5. A. D. Brjuno (1970), *Instability in a Hamiltonian system and the distribution of asteroids*, Math USSR-Sb. **12**, 271–312.
6. C. R. Chapman, J. G. Williams, and W. K. Hartman (1978), *The asteroids*, Annual Rev. Astron. Astrophys. **16**, 33–75.
7. J. Hadamard (1898), *Les surfaces à courbure opposées et leurs lignes géodésiques*, J. Math. Pures Appl. (5) **4**, 27–73.
8. J. E. Littlewood (1952), *On the problem of n bodies*, Comm. Sem. Math. Univ. Lund, Suppl. Tome (Marcel Riesz volume), 143–151.
9. J. Koiller, J. M. Balthazar, and T. Yokoyama (1987), *Relaxation—chaos phenomena in celestial mechanics* I. *On Wisdom's model for the 3/1 Kirkwood gap*, Physica D **26**, 85–122.
10. J. C. Lagarias (1985), *The 3x + 1 problem and its generalizations*, Amer. Math. Monthly **92**, 3–23.
11. J. Laskar (1990), *The chaotic motion of the solar system: a numerical estimate of the size of the chaotic zones*, Icarus **88**, 266–291.
12. H. Minkowski (1904), *Zur geometrie der Zahlen*, Gesammelte Abhandlungen, Vol. II, pp. 43–52.
13. M. Morse and G. Hedlund (1938), *Symbolic dynamics*, Amer. J. Math. **60**, 815–866.
14. S. J. Peale (1976), *Orbital resonances in the solar system*, Annual Rev. Astron. Astrophys. **14**, 215–246.
15. H. Poincaré (1885), *Sur les courbes définies par les équations différentielles*, J. Math. Pures Appl. (4) **1**, 167–244.
16. H. Poincaré (1912), *Sur une theoreme de geometrie*, Rend. Circ. Mat. Palermo **33**, 375–407.

17. S. Smale (1967), *Differentiable dynamical systems*, Bull. Amer. Math. Soc. **73**, 747–817.
18. G. J. Sussman and J. Wisdom (1988), *Numerical evidence that the motion of Pluto is chaotic*, Science **241**, 433–437.
19. J. Wisdom (1982), *The origin of the Kirkwood gaps: a mapping for asteroidal motion near the* 3/7 *acommensurability*, Astron. J. **87**, 577–593.
20. J. Wisdom (1987), *Chaotic behaviour in the solar system*, Proc. Roy. Soc. London A **413**, 109–129.

§2. CONTINUED FRACTIONS AS DYNAMICAL SYSTEMS

1. P. Arnoux and A. Nogueira (1991), *Mesures de Gauss pour des algorithmes de fractions continues multidimensionnelles*, preprint, Univ. of Paris VII.
2. L. R. Goldberg and C. Tresser (1991), *Rotation orbits and the Farey tree*, preprint.
3. A. Ya. Khinchin (1935), *Metrische Kettenbruchprobleme*, Compositio Math. **1**, 359–382.
4. A. Ya. Khinchin (1936), *Zur metrischen Kettenbruchtheorie*, Compositio Math. **3**, 276–285.
5. A. Ya. Khinchin (1964), *Continued fractions*, Univ. of Chicago Press, Chicago, IL.
6. J. R. Kinney (1960), *Note on a singular function of Minkowski*, Proc. Amer. Math. Soc. **11**, 788–794.
7. J. C. Lagarias and Y. Peres (1992), *The Farey shift and Minkowski's ?-function*, in preparation.
8. P. Lévy (1936), *Sur le development en fraction continue d'un nombre choisi au hasard*, Compositio Math. **3** 286–303.
9. D. Meyer (1990), *Continued fractions and related transformations*, Ergodic Theory, Symbolic Dynamics and Hyperbolic Spaces (T. Bedford, M. Keane, and C. Series, Eds.), Oxford Univ. Press, Oxford, pp. 175–222.
10. W. Parry (1962), *Ergodic properties of some permutation processes*, Biometrika **49**, 151–154.
11. K. Petersen (1983), *Ergodic theory*, Cambridge Univ. Press.
12. W. Philipp (1967), *Some metrical theorems in number theory*, Pacific J. Math. **20**, 109–127.
13. M. Pollicott (1986), *Distribution of closed geodesics on the modular surface and continued fractions*, Bull. Math. Soc. France **114**, 431–446.
14. I. Richards (1981), *Continued fractions without tears*, Math. Mag. **54**, 163–171.
15. R. Salem (1943), *Singular monotone functions*, Trans. Amer. Math. Soc. **53**, 427–439.
16. C. Series (1982), *Non-Euclidean geometry, continued fractions, and ergodic theory*, Math. Intelligencer **4**, 24–31.
17. C. Series (1985a), *The geometry of Markoff numbers*, Math. Intelligencer **7**, 20–29.
18. C. Series (1985b), *The modular surface and continued fractions*, J. London Math. Soc. **31**, 69–80.
19. E. Wirsing (1974), *On the theorem of Gauss-Kuzmin-Levy and a Frobenius type theorem for function spaces*, Acta Arith. **24**, 507–528.

§3. INTEGRABLE HAMILTONIAN DYNAMICAL SYSTEMS

1. D. Bayer and J. C. Lagarias (1989), *The nonlinear geometry of linear programming II. Legendre transform coordinates and central trajectories*, Trans. Amer. Math. Soc. **314**, 527–581.
2. P. A. M. Dirac (1950), *Generalized Hamiltonian dynamics*, Canad. Math. J. **2**, 129–148.
3. L. D. Faddeev and L. A. Takhtajan (1987), *Hamiltonian methods in the theory of solitons*, Springer-Verlag, New York.
4. H. Flaschka (1974), *The Toda lattice. I*, Phys. Rev. B **9**, 1924–1925.
5. H. Ito (1991), *Action-angle coordinates at singularities for analytic integrable systems*, Math. Z. **206**, 363–407.
6. J. Moser (1975), *Finitely many mass points on the line under the influence of an exponential potential—an integrable system*, Lecture Notes in Phys., vol. 38, Springer-Verlag, Berlin and New York, pp. 467–497.

7. J. Moser (1980), *Various aspects of integrable Hamiltonian systems*, Dynamical Systems, CIME Lectures, Birkhäuser, Boston, MA, pp. 233–290.

8. I. Percival (1986), *Integrable and nonintegrable Hamiltonian systems*, Nonlinear Dynamics Aspects of Particle Accelerators, Lecture Notes in Phys., vol. 247, Springer-Verlag, New York, pp. 12–36.

9. J. Pöschel (1989), *On elliptic lower dimensional tori in Hamiltonian systems*, Math. Z. **202**, 559–608.

10. W. W. Symes (1982), *The QR-algorithm and scattering for the finite non-periodic Toda lattice*, Phys. D **4**, 275–280.

§4. KAM THEORY AND SMALL DIVISORS

1. V. I. Arnol'd (1963), *Small denominators and problems of stability of motion in classical and celestial mechanics*, Russian Math. Surveys **18**, 85–191.

2. J. Bellissard (1985), *Small divisors in quantum mechanics*, Proc. NATO ASI Workshop: Chaotic Behavior in Quantum Systems (G. Casati, ed.), Plenum, New York, pp. 11–35.

3. J. M. Greene (1979), *A method for determining a stochastic transition*, J. Math. Phys. **20**, 1183–1201.

4. A. Kolmogorov (1954), *On conservation of conditionally periodic motions under small perturbations of the Hamiltonian*, Dokl. Akad. Nauk. SSR **98**, 527–530.

5. J. Mather (1982), *Existence of quasiperiodic orbits for twist mappings of the annulus*, Topology **21**, 457–467.

6. J. Mather (1984), *Non-existence of invariant circles*, Ergodic Theory Dynamical Systems **4**, 301–309.

7. R. S. MacKay (1985), *Converse KAM: theory and practice*, Comm. Math. Phys. **98**, 469–512.

8. R. S. MacKay (1986), *Transition to chaos for area-preserving maps*, Nonlinear Dynamics Aspects of Particle Accelerators, Lecture Notes in Phys., vol. 247, Springer-Verlag, New York, pp. 390–454.

9. R. S. MacKay, J. D. Meiss, and I. C. Percival (1984), *Transport in Hamiltonian systems*, Phys. D **13**, 55–81.

10. J. Moser (1973), *Stable and random motion in dynamical systems*, Princeton Univ. Press, Princeton, NJ.

11. H. Rüssmann (1989), *Nondegeneracy in the perturbation theory of dynamical systems*, Number Theory and Dynamical Systems (M. M. Dodson and J. M. Vickers, ed.), Cambridge Univ. Press, pp. 5–19.

§5. ITERATION OF ANALYTIC FUNCTIONS AND SMALL DIVISORS

1. I. N. Baker and P. J. Rippon (1984), *Iteration of exponential functions*, Ann. Acad. Sci. Fenn. Ser. A I Math. **9**, 49–77.

2. A. Beardon (1991), *Iteration of rational functions*, Springer-Verlag, New York.

3. A. D. Brjuno (1971), *Analytic form of differential equations*, Trans. Moscow Math. Soc. **25**, 131–288.

4. M. R. Herman (1987), *Recent results and some open questions on Siegel's linearization theorem on germs of complex analytic diffeomorphisms near a fixed point*, VIIth Internat. Congress on Math. Physics (Marseille 1986), World Scientific, pp. 138–184.

5. J. Martinet (1981), *Normalisation des champs le vecteurs, d'apres A. D. Brjuno*, Séminaire Bourlaki Exp. 582, Lecture Notes in Math., vol. 901, Springer-Verlag, New York, pp. 55–70.

6. E. Schröder (1870), *Ueber unendliche viele Algorithmen zur Aüflosung der Gleichungen*, Math. Ann. **2**, 317–365.

7. E. Schröder (1871), *Ueber iterirte Functionen*, Math. Ann. **3**, 296–322.

8. C. L. Siegel (1942), *Iteration of analytic functions*, Ann. of Math. **43**, 807–812.

9. J.-C. Yoccoz (1988), *Linéarisation des germes de diffeomorphisms holomorphes de* $(\mathbb{C}, \mathbf{0})$, C. R. Acad. Sci. Paris Sér. I. Math. **306**, 55–58.

§6. Mode-locking

1. V. I. Arnol'd (1965), *Small denominators* I. *Mappings of the circumference onto itself*, Amer. Math. Soc. Transl. Ser. 2, vol. **46**, Amer. Math. Soc., Providence, RI, pp. 213–284.

2. V. I. Arnol'd (1983), *Remarks on perturbation theory for problems of Mathieu type*, Russian Math. Surveys **38**, 215–233.

3. P. Bak, T. Bohr, and M. Høgh-Jensen (1985), *Mode-locking and the transition to chaos in dissipative systems*, Phys. Scripta **9**, 50–58.

4. K. M. Brucks and C. Tresser (1991), *A Farey tree organization for locking regions for simple circle maps*, preprint.

5. P. Cvitanović, M. H. Jensen, L. P. Kadanoff, and I. Procaccia (1985), *Renormalization, unstable manifolds and the fractal structure of mode locking*, Phys. Rev. Lett. **55**, 343–346.

6. P. Cvitanović, B. Shraiman, and B. Söderberg (1985), *Scaling laws for mode locking in circle maps*, Phys. Scripta **32**, 263–270.

7. M. J. Feigenbaum (1988), *Presentation functions, fixed points, and a theory of scaling function dynamics*, J. Stat. Phys. **52**, 527–569.

8. M. R. Herman (1977), *Mesure de Lebesgue et nombre de rotation*, Geometry and Topology, Lecture Notes in Math., vol. 597, Springer-Verlag, New York, pp. 271–293.

9. M. Høgh-Jensen, P. Bak, and T. Bohr (1983), *Complete Devil's staircase, fractal dimension and universality of mode-locking structure in the circle map*, Phys. Rev. Lett. **50**, 1637–1639.

10. M. V. Jakobson (1988), *Onset of stochasticity for some one-dimensional systems*, Physica D **33**, 157–164.

11. L. Jonker (1990), *The scaling of Arnol'd tongues at degenerate fold bifurcations*, Physica D **41** , 275–281.

12. I. Kan and J. A. Yorke (1990), *Antimonotonicity: concurrent creation and annihilation of periodic orbits*, Bull. Amer. Math. Soc. **23**, 469–476.

13. K. M. Khanin (1991), *Universal estimates for critical circle maps*, Chaos (to appear).

14. S. Kim and S. Ostlund (1985), *Renormalization of mappings of the two-torus*, Phys. Rev. Lett. **55**, 1165–1168.

15. S. Kim and S. Ostlund (1986), *Simultaneous rational approximation in the study of dynamical systems*, Phys. Rev. A **34**, 3426–3434.

16. S. Kim and S. Ostlund (1989), *Universal scaling in circle maps*, Physica D **39**, 365–392.

17. R. S. MacKay (1991), *Scaling exponents at the transition by breaking of analyticity for incommensurate structures*, Physica D **50**, 71–79.

18. S. Ostlund, D. Rand, J. Sethna, and E. Siggia (1983), *Universal properties of the transition from quasi-periodicity to chaos in dissipative systems*, Physica **80**, 303–342.

19. S. Ostlund and S.-H. Kim (1985), *Renormalization of quasiperiodic mappings*, Phys. Scripta **9**, 193–198.

20. I. Procaccia, S. Thomae, and C. Tresser (1987), *First-return maps as a unified renormalization scheme for dynamical systems*, Phys. Rev. A **35**, 1884–1990.

21. B. Simon (1982), *The rotation number for almost periodic potentials*, Comm. Math. Phys. **84**, 403–438.

22. Ya. G. Sinai and K. M. Khanin (1988), *Renormalization group method in the theory of dynamical systems*, Internat J. Modern Phys. B **2** 147–165.

23. D. Sullivan (1989), *Bounded structure in infinitely renormalizable mappings*, Universality in Chaos (P. Cvitanović, ed.), 2nd ed, Adam Hilger, Bristol.

24. G. Swiatek (1988), *Rational rotation numbers for maps of the circle*, Comm. Math. Phys. **119**, 109–128.

§7. Quasicrystals

1. J. P. Allouche and M. Mendes-France (1986), *Finite automata and zero temperature quasicrystal Ising chain*, J. Physique **47**, C3-63–C3-73.

2. S. Aubry and P. le Daeron (1983), *The discrete Frankel-Kontorova model*, Physica D **8**, 381–422.

3. J. Bellissard, R. Lima, and D. Testard (1985), *Almost periodic Schrödinger operators*, Mathematical Physics, Lectures on Recent Results (L. Streit, ed.), World Scientific, Singapore pp. 1–64.

4. E. Bombieri and J. E. Taylor (1987), *Quasicrystals, tilings and algebraic number theory: some preliminary connections*, Contemp. Math., vol. 64, Amer. Math. Soc., Providence, RI, pp. 241–264.

5. N. G. de Bruijn (1973), *A theory of generalized functions, with applications to Wigner distribution and Weyl correspondence*, Nieuw Arch. Wisk. **21**, 205–280.

6. N. G. de Bruijn (1981), *Algebraic theory of Penrose's non-periodic tilings of the plane*. I, Proc. Konink. Ned. Akad. Weknsch., vol. 43, 39–52; II, 53–66.

7. N. G. de Brujn (1986), *Quasicrystals and their Fourier transform*, Indag. Math. **48**, 123–152.

8. S. E. Burkov (1987), *One-dimensional model of the quasicrystalline alloy*, J. Stat. Phys. **47**, 409–438.

9. J. W. Cahn and J. E. Taylor (1987), *An introduction to quasicrystals*, Contemp. Math., vol. 64, Amer. Math. Soc., Providence, RI, pp. 265–286.

10. E. M. Coven and M. Keane (1971), *The structure of substitution minimal sets*, Trans. Amer. Math. Soc. **162**, 89–102.

11. F. M. Dekking (1981), *On the structure of self-generating sequences*, Seminar on Number Theory—Bordeaux 1980-81 (Talence 1980-81), Exp. No. 31, Univ. Bordeaux I, 6 pp.

12. F. Gähler (1986), *Some mathematical problems arising in the study of quasicrystals*, J. Physique **47**, 115–123.

13. T. Janssen and A. Janner (1987), *Incommensurability in crystals*, Adv. Physics **36**, 519–624.

14. A. Katz and M. Duneau (1986), *Quasiperiodic patterns and icosahedral symmetry*, J. Physique **47**, 181–196.

15. W. Kolakowski (1966), *Self-generating runs, Problem* 5304 *, Solution*, American Math. Monthly **73**, 681–682.

16. M. Keane (1990), *Ergodic theory and subshifts of finite type*, Ergodic Theory, Symbolic Dynamics and Hyperbolic spaces (T. Bedford, et al., eds.), Oxford Univ. Press, Oxford, pp. 35–70.

17. A. R. Kortan, R. S. Becker, F. A. Thiel, and H. S. Chen (1990), *Real-space atomic structure of a two-dimensional decagonal quasicrystal*, Phys. Rev. Lett. **64**, 200–203.

18. D. Levine and P. J. Steinhardt (1986), *Quasicrystals* I. *Definition and Structure*, Phys. Rev. B **34**, 596–616.

19. J. C. Martin (1973), *Minimal flows arising from substitutions of non-constant length*, Math. Systems Theory **7**, 73–82.

20. M. Mendes-France and A. von der Poorten (1981), *Arithmetic and analytic theory of paper folding sequences*, Bull. Austral. Math. Soc. **24**, 123–130.

21. Y. Meyer (1972), *Algebraic numbers and harmonic analysis*, North-Holland, Amsterdam.

22. J. Milnor (1976), *Hilbert's problem* 18: *On crystallographic groups, fundamental domains and on sphere packing*, Proc. Sympos. Pure Math., vol. 28, Amer. Math. Soc., Providence, RI, pp. 491–506.

23. G. Y. Onoda, P. J. Steinhardt, D. P. DiVicennzo, and J. E. Socolar (1988), *Growing perfect quasicrystals*, Phys. Rev. Lett. **60**, 2653–2656.

24. R. Penrose (1989), *Tilings and quasicrystals: a non-local growth problem?* Introduction to the Mathematics of Quasicrystals (M. Jaric, ed.), Academic Press, New York, pp. 53–79.

25. R. Porter (1988), *The applications of the properties of Fourier transforms to quasicrystals*, M.Sc. Thesis, Rutgers Univ.

26. F. Queffélec (1987), *Substitution dynamical systems-spectral analysis*, Lecture Notes in Math., vol. 1294, Springer-Verlag, New York.

27. C. Radin (1986), *Crystals and quasicrystals: a continuum model*, Comm. Math. Phys. **105**, 385–390.

28. C. Radin (1991), *Global order from local sources*, Bull. Amer. Math. Soc. **25**, 335–364.

29. G. Rauzy (1982), *Nombres algébriques et substitutions*, Bull. Soc. Math. France **110**, 147–178.

30. R. Salem (1963), *Algebraic numbers and Fourier analysis*, Heath, Boston, MA. (Reprint: Wadsworth, Belmont, CA.)

31. D. Shechtman, I. Blech, P. Gratias, and J. W. Cahn (1984), *Metallic phase with long-range orientational order and no translational symmetry*, Phys. Rev. Lett. **53**, 1951–1954.

32. M. Senechal and J. Taylor (1990), *Quasicrystals: The view from les Houches*, Math. Intelligencer **12**, 54–64.

33. J. O. Shallit (1992), *Real numbers with bounded partial quotients: a survey*, Enseign. Math. (to appear).

34. B. Simon (1982), *Almost periodic Schrödinger operators: a review*, Adv. Appl. Math. **3**, 463–490.

35. B. Simon (1984), *Fifteen problems in mathematical physics*, in: Perspectives in Mathematics, Birkhäuser-Verlag, Basel, pp. 423–454.

36. J. E. Socolar and P. J. Steinhardt (1986), *Quasicrystals II. Unit-cell configurations*, Phys. Rev. B **34**, 617–647.

37. A. Thue (1912), *Uber die Gegenseitige Lage Gleicher Teile Gewisser Zahlenreihen*, Kristiania (Oslo) Viadens. Skrifter I, Math.-Naturwiss. Kl., No. 1.

38. S. J. L. van Eijndhoven (1987), *Functional-analytic characterizations of the Gelfand-Shilov spaces $S(a, b)$* , Indag. Math. **90**, 133–144.

AT&T BELL LABORATORIES

MURRAY HILL, NEW JERSEY 07974
E-mail address: jcl@research.att.com

Proceedings of Symposia in Applied Mathematics
Volume 46, 1992

The Mathematics of Random Number Generators

George Marsaglia
Supercomputer Computations Research Institute
and
Department of Statistics
The Florida State University

1 Introduction

This article is based on a handout for the AMS 1991 Summer Short Course in Orono Maine. Its title is not meant to be presumptuous, nor its content all-inclusive. It should be taken in context with the theme of the Summer Course: The unreasonable effectiveness of number theory. Most of the mathematics behind random number generators comes from number theory, so, in a sense, number theory *is* the mathematics of random number generators. We discuss some of it here. For results already in the literature, a brief description of the role of number theory is given, and a few proofs sketched. More detail is given for the mathematics of some promising new kinds of generators.

A **random number generator** is a computer procedure that scrambles the bits of a current number or set of numbers to produce a new number, in such a way that the result appears to be randomly distributed among the set of possible numbers and independent of the previously generated numbers. As experiments over the years have shown, this appears surprisingly easy to do. A wide variety of scrambling methods have been proposed. Random number generators are provided for most computer systems or software packages, and they work remarkably well—at least for limited use when only a few hundreds or thousands of numbers are required. But experience with very fast computers doing Monte Carlo problems requiring samples of hundreds of millions or billions has shown that the random number generator should be carefully chosen.

The most commonly used bit-scrambling method uses multiplication. Here is an example, using digits of the more familiar base 10, rather than the bits for base 2 that are used in computers: the current 'random number' of, say, 10 digits is multiplied by a constant then the last 10 digits of the product are taken as the new random number. For this example, start with an initial random number (the seed), say $x = 5362817283$, multiply by a constant, say $a = 81734027$ to get

1991 *Mathematics Subject Classification*. Primary 65C10, 10A30.
This paper is in final form and no version of it will be submitted for publication elsewhere.

© 1992 American Mathematical Society
0160-7634/92 $1.00 + $.25 per page

an 18 digit product: $ax = 435141161293554841$. Then take the last 10 digits of that product as the new random number: $x = 1293554841$. For the next x, form the product $ax = 105727446300274707$, take the last 10 digits (reduce modulo 10^{10}) to get the new $x = 6300274707$, and so on. This is called a congruential random number generator. With proper choice of multiplier and modulus, such a generator produces a sequence of numbers that are difficult to distinguish from truly random numbers. A good congruential generator could be used to run the casinos in Las Vegas and Atlantic City and all the state lotteries with no one the wiser except those in the know.

The first electronic computers of the late 1940's used random number generators much like the one above, and the intervening forty years have seen many arithmetic or algebraic schemes that seem to produce randomness, even though the results are completely deterministic. But modern computer speeds and exotic architectures make possible massive Monte Carlo simulations for which standard generators may not be suitable. A table giving some of the most common old, as well as promising new, random number generators is given at the end of this handout.

2   Underlying Theory

Virtually all random number generators are based on theory which may be described as follows: We have a finite set X and a function $f : X \rightarrow X$ that takes elements of X into other elements of X. Given an initial (seed) value, $x \in X$, (x might be a single computer word or a vector of computer words), the generated sequence is

$$x, f(x), f^2(x), f^3(x), \ldots,$$

where $f^2(x)$ means $f(f(x))$, $f^3(x)$ means $f(f^2(x)) = f(f(f(x)))$ and so on. The three most common classes of random number generators are 1) Congruential, 2) Shift-register and 3) Lagged-Fibonacci.

2.1   Congruential generators

For **congruential generators**, the finite set X is the set of reduced residues of some modulus m and $f(x) = ax + b \bmod m$. Thus, with an initial element $x_0 \in X$, the generated sequence is

$$x_0, x_1, x_2, \ldots \qquad \text{with } x_{n+1} = ax_n + b \bmod m.$$

A wide variety of choices for a, b and m have been described in the literature; see, particularly, Marsaglia [3] or Knuth [1] for methods for finding periods and establishing structure of congruential sequences. With periods around 2^{32}, congruential generators have been used successfully in Monte Carlo simulations for the past 35 years. Most of the random number generators provided by computer systems or software packages are congruential generators. But random points in higher dimensions with coordinates produced by congruential generators show a crystalline regularity that makes them unsuitable for certain applications, [2] and that, taken with their relatively short periods, has led to the gradual adoption of longer-period generators for serious Monte Carlo studies.

Every congruential sequence of reduced residues of a modulus m:

$$x_0, x_1, x_2, x_3, \ldots, \qquad x_n = ax_{n-1} + b \bmod m$$

may be transformed to the *fundamental sequence*

$$y_0, y_1, y_2, \ldots, \qquad y_n = 1 + a + \cdots + a^{n-1} \bmod m.$$

Proof: $x_{i+1} = vy_i + x_0$, where $v = x_0(a-1) + b$.

Thus the period and structure of the general x sequence may be established from that of the fundamental y sequence. The period of the sequence $y_n \bmod m$ is that of $y_n \bmod m/d$, where $d = \gcd[m, x_0(a-1) + b]$. Sequences in which $\gcd(a, m) > 1$ are the most important from the standpoint of applications, and then it is the *effective period* of the $y_n \bmod m$ that plays a key role. The effective period is the order of a for modulus m: the least integer t such that $a^t = 1 \bmod m$. Then the fundamental sequence $y_n \bmod m$ is made up of a block $\{B\} = \{y_0, y_1, \ldots, y_{t-1}\}$ of t distinct residues of m, followed by translates of that block:

$$\{B\}, \{B+c\}, \{B+2c\}, \{B+3c\}, \ldots,$$

where $c = y_t$. For most applications, translates of the block $\{B\}$ cannot reasonably be considered independent sets. In that sense, only t successive elements of the congruential sequence are useful—hence the term *effective* period of a congruential generator. See Marsaglia [3] for details.

2.2 Shift-Register Generators

For **shift-register generators**, the underlying theory has the finite set X as the set of $1 \times n$ binary vectors $x = (b_1, b_2, \ldots, b_n)$ and the function f is a linear transformation, $f(x) = xT$, with T an $n \times n$ binary matrix and all arithmetic mod 2. With an initial binary vector x the sequence is

$$x, xT, xT^2, xT^3, \ldots$$

with the matrix T chosen so that the period is long and multiplication by T is reasonably fast in computer implementation. The most common application has T of the form $T = (I + R^a)(I + L^b)$, where R and L are matrices that shift the components of a vector right or left by one position:

$$(b_1, b_2, \ldots, b_n)R = (0, b_1, b_2, \ldots, b_{n-1}), \qquad (b_1, b_2, \ldots, b_n)L = (b_2, b_3, \ldots, b_n, 0),$$

Thus the product βT can be formed be means of right and left shifts and exclusive-or's of computer words viewed as binary vectors. Shift-register generators are sometimes called Tausworthe generators.

The use of shift-register generators is declining. Those based on standard computer words, $n = 32$, do not perform as well on tests of randomness as do congruential generators, and their periods, like those of congruential generators, are too short. Their main use is in forming part of a combination generator. The use of shift-register generators with extremely long binary vectors, $n = 607, 1279$ or even 9689 still has some attraction, for the periods can be made extremely long, $2^n - 1$, and special hardware (called shift registers and hence the name of the general method) is easily constructed for their implementation. They are mainly used in special purpose machines for Monte Carlo studies in Physics.

Full-period shift-register generators are usually called for in applications—that is, sequences that include all $2^n - 1$ possible non-null binary vectors. We sketch proof of a condition that characterizes full-period sequences. It also plays a role in establishing the periods of the more important lagged-Fibonacci generators:

Theorem *In order that the sequence*

$$\beta, \beta T, \beta T^2, \ldots$$

have period $2^n - 1$ for every non-null binary seed vector β, it is necessary and sufficient that the $n \times n$ binary matrix T have order $2^n - 1$ in the group of $n \times n$ nonsingular binary matrices.

Outline of a sufficiency proof:

1. Assume T has order $t = 2^n - 1$. Consider the set of distinct powers of T:
   $$\mathcal{S} = \{I, T, \ldots, T^{t-1}\}.$$

2. Let $\mathcal{P}$ be the set of non-null polynomials in T of degree less than n with coefficients in the base field $\{0, 1\}$.

3. Both $\mathcal{S}$ and $\mathcal{P}$ have $2^n - 1$ elements. Indeed, $\mathcal{S} = \mathcal{P}$, because T satisfies its characteristic equation and hence every element T^j in $\mathcal{S}$ equals, through Euclid's algorithm for polynomials, an element of $\mathcal{P}$.

4. Since the elements of $\mathcal{S}$ are, by assumption, nonsingular and distinct, so are those of $\mathcal{P}$.

5. If the period of the sequence were less than $t = 2^n - 1$, then $\beta T^j = \beta$ for some non-null β and $j < t$. This requires that the matrix $T^j - I$, non-null and in $\mathcal{P}$ through reduction, be singular. But the polynomials in $\mathcal{P}$ are those of $\mathcal{S}$, and all are nonsingular.

Proof of necessity:

1. If $\beta, \beta T, \beta T^2, \ldots$ has period $t = 2^n - 1$ for any non-null initial binary vector β, then the order of T cannot be less than t.

2. Because $\gamma T^t = \gamma$ for every $1 \times n$ binary vector γ, the null space of the $n \times n$ matrix $T^t - I$ has dimension n, so $T^t = I$ and the order of T cannot be greater than t.

Note that the above proof applies for matrices with elements in the field of residues mod p for any prime p, not only $p = 2$. This provides a criterion for establishing the periods of **extended congruential generators,** in which each new integer is a linear combination of the previous k integers:

$$x_{n+1} = a_1 x_{n-1} + a_2 x_{n-2} + \cdots + a_k x_{n-k}.$$

The underlying theory for extended congruential generators has X as the set of $1 \times k$ vectors and the iterating function f a linear transformation given by a $k \times k$ matrix T with elements reduced residues of the prime p:

$$x, xT, xT^2, xT^3, \ldots.$$

For applications where integers are generated one at a time, the matrix T is a companion matrix, so that

$$xT = (x_1, x_2, \ldots, x_k)T = (x_2, x_3, \ldots, x_k, a_1 x_k + a_2 x_{k-1} + \cdots + a_k x_1).$$

If T is a general $k \times k$ matrix over the field mod p then one may generate vectors of k integers at a time by this iteration. The modulo p version of the above theorem holds: $x, xT, xT^2, \ldots$ *has period* $p^k - 1$ *for every non-null* $1 \times k$ *initial vector* x *if, and only if, the order of the matrix* T *is* $p^k - 1$ *in the group of nonsingular matrices modulo the prime* p.

Finding such T's leads to interesting searches and uses of number theory. Unfortunately, arithmetic modulo a prime p is quite slow in most computers, and advocates of such generators have not convinced many potential users that k multiplications and additions modulo p for each new number are worth the effort. There are more efficient ways to produce long-period sequences.

2.3 Lagged-Fibonacci Generators

For **lagged-Fibonacci generators,** the finite set X is the set of $1 \times r$ vectors $x = (x_1, x_2, \ldots, x_r)$ with elements x_i in some finite set S on which there is a binary operation $\diamond$. The function f is defined by

$$f(x_1, x_2, \ldots, x_r) = (x_2, x_3, x_4, \ldots, x_r, x_1 \diamond x_{r+1-s}).$$

Informally, a lagged-Fibonacci sequence is described by means of a set of r seed values followed by the rule for generating succeeding values:

$$x_1, x_2, \ldots, x_r, x_{r+1}, \ldots \qquad \text{with } x_n = x_{n-r} \diamond x_{n-s},$$

but to formally define and establish the period and structure of such sequences they must be viewed as iterates $x, f(x), f^2(x), \ldots$ on the set X of $1 \times r$ vectors with elements in the set S on which the binary operation $\diamond$ is defined.

Various choices for S and $\diamond$ lead to interesting sequences—for example, when S is the set of reduced residues of some modulus m and $\diamond$ is addition or subtraction mod m; S is the set of reduced residues relatively prime to m and $\diamond$ is multiplication; S is the set of $1 \times k$ binary vectors and $\diamond$ is addition of binary vectors (exclusive-or); S is the set of floating-point computer numbers $0 \leq x < 1$ having 24-bit fractions and $x \diamond y = \{$if $x > y$ then $x - y$ else $x - y + 1\}$. Such generators are often designated $F(r,s,\diamond)$ generators.

Implementations of lagged-Fibonacci generators require a table of the previous r numbers, say $L(1), L(2), \ldots, L(r)$ and two pointers I,J pointing to the last values used in the previous $x \diamond y$ operation. Then instructions equivalent to these are programmed:

$K \leftarrow L[I] \diamond L[J]$
$L[I] \leftarrow K$
$I \leftarrow I - 1$: **if** $I = 0$ **then** $I \leftarrow r$
$J \leftarrow J - 1$: **if** $J = 0$ **then** $J \leftarrow r$
return K

Here is how the periods of the above lagged-Fibonacci generators may be established. First, the $F(r, s, - \bmod 2^n)$ generator described informally by $x_n = x_{n-r} - x_{n-s} \bmod 2^n$. The underlying theory calls for a finite set X of $1 \times r$ vectors of integers mod 2^n and an $r \times r$ companion matrix T such that

$$(x_1, x_2, x_3, \ldots, x_r)T = (x_2, x_3, \ldots, x_r, x_{r+1-s} - x_1 \bmod 2^n).$$

The idea is to build up the period from mod 2 to mod 2^2, mod 2^3, and so on. The above theorem shows that T must have order $t = 2^r - 1$ for modulus 2. One then checks that the order of T is $2t$ for modulus 4 and $4t$ for modulus 8. It then follows that the period for modulus 2^n is $2^{n-1}(2^r - 1)$. See Marsaglia and Tsay [4].

The maximal period for lagged-Fibonacci generators $F(r, s, \star \bmod 2^n)$ using multiplication is $2^{n-3}(2^r - 1)$, which may be established for certain r, s by expressing the residues modulo 2^n as the direct product of the cyclic group generated by -1 and the cyclic group generated by 3. Every residue may be represented as $(-1)^a 3^b$, and multiplication reduces to addition on exponents, so that the theory for $F(r, s, + \bmod 2^k)$ sequences applies.

For lagged-Fibonacci generators using the exclusive-or operation $\oplus$, the maximal period is only $2^r - 1$, far short of the nearly 2^{n+r} attainable for those using addition, subtraction or multiplication modulo 2^n. The maximal $F(r, s, \oplus)$ periods may be established directly by the above theorem for modulus 2.

While examples of generators of each of the three standard methods described above are widely used and—for most purposes—work quite well, new methods are always being developed. All standard generators (with the exception of lagged-Fibonacci using multiplication) fail one or more stringent tests of randomness such as those described in Marsaglia [5], and many of them have periods too short for the huge samples that current computer speeds make possible.

3 Periods and Seed Values

An ideal generator should have period as great as the number of possible choices for seed values. Then, if the seed values are $x_1, x_2, \ldots, x_r$ and the sequence is strictly periodic, every possible r-tuple of x's will appear in the full sequence—a desirable uniformity property. Except for trivial cases of little interest, the lagged-Fibonacci generators—until recently the record holders for long periods—do not have this property. The lagged-Fibonacci generators $F(r,s,- \bmod 2^{32})$, $F(r,s,* \bmod 2^{32})$ or $F(r,s,- \bmod 1)$ have periods on the order of $2^{32+r}, 2^{30+r}, 2^{24+r}$, far short of the ideals of $2^{32r}, 2^{30r}$ or 2^{24r} that are the number of possible choices of seed values. (Nonetheless, their periods are still far longer than those for $F(r,s,\oplus)$ generators using exclusive-or, for which the period is at most 2^r, whatever the word size.)

Of course, congruential generators satisfy the above criterion: the period equals the number of choices for the seed value. But that period, on the order of 2^{32}, is far too short for modern needs. The full period can be quickly exhausted and it cannot provide the variety of possible k-tuples of numbers that probability theory says should be encountered in long streams.

There exist longer period generators for which the period equals the number of choices of seed values. One such class is the extended congruential generators mentioned above for a prime modulus. For example, if p is the prime $2^{31}-1$ then the sequence produced by $x_n = 1999x_{n-1}+4444x_{n-2} \bmod p$ has period p^2-1 for any initial seed values x_1, x_2 not both zero, and there are p^2-1 possible choices. One can find constants, c_1, c_2, c_3 so that $x_n = c_1x_{n-1} + c_2x_{n-2} + c_3x_{n-3} \bmod p$ has period $p^3 - 1$ for any of the $p^3 - 1$ possible seeds x_1, x_2, x_3 not all zero, and so on: for any prime p, and any lag k, there are k constants such that the sequence $x_n = c_1x_{n-1} + \cdots c_kx_{n-k} \bmod p$ has period $p^k - 1$ for any choice of k seed values not all zero. This may be established by showing the iterating matrix T has order $p^k - 1$ in the group of nonsingular matrices modulo p.

4 A New Class of Generators

The rest of this handout is devoted to description of a new class of random number generators. They have astonishingly long periods and are based on ordinary number theory. These new generators, developed by Marsaglia and Zaman, are described in [6]. They are like lagged-Fibonacci generators using addition or subtraction of the residues of some modulus, but with each addition or subtraction a "carry" or "borrow" is noted and used in the succeeding operation.

4.1 The New Class: add-with-carry generators

We introduce add-with-carry generators with a simple example. Consider the classical Fibonacci sequence

$$0, 1, 1, 2, 3, 5, 8, 13, 21, 34, 55, 89, \ldots,$$

with each element the sum of the previous two. If we take this sequence mod 10, we have an example of a lagged-Fibonacci sequence with lags $r = 2$ and $s = 1$ and binary operation $v \diamond w = v + w \bmod 10$:

$$0, 1, 2, 2, 3, 5, 8, 3, 1, 4, 5, 9, 4, 3, 7, \ldots.$$

The informal description of the sequence is $x_n = x_{n-2} + x_{n-1} \bmod 10$, but to formally describe it and define and establish its period we need the finite set X of 1×2 vectors $x = (x_1, x_2)$ with elements reduced residues of 10 and the iterating function f defined by $f[(x_1, x_2)] = (x_2, x_1 + x_2 \bmod m)$. Since f has an inverse, for any initial vector $x \in X$ the sequence

$$x, f(x), f^2(x), f^3(x), \ldots$$

is strictly periodic. Depending on the initial vector x, there is a longest cycle of period 60 and shorter cycles of periods $1, 3, 4, 12$ and 20. Each period is the lcm of the periods for moduli 2 and 5. Figure 1 gave the directed graph of this generator.

Now consider the add-with-carry version of this generator. We assign two initial values, again $0, 1$, but also an initial "carry bit", say 0. Then each new digit is the sum of the previous two digits *plus the carry bit*. The result is taken mod 10 and the next carry bit set to 1 or 0 according to whether or not the sum exceeds 10. Using a superscript to indicate the carry bit, the sequence of digits becomes

$$0, 1^0, 1^0, 2^0, 3^0, 5^0, 8^0, 3^1, 2^1, 6^0, 8^0, 4^1, 3^1, 8^0, 1^1, 0^1, 2^0, \ldots.$$

Formally, as before, we have a sequence of iterates $x, f(x), f^2(x), \ldots$. But now our x's come from the set X of 1×3 vectors $x = (x_1, x_2, c)$ with x_1, x_2 reduced residues of 10 and c the "carry bit", 0 or 1. Then the iterating function f is

$$f(x_1, x_2, c) = \begin{cases} (x_2, x_1 + x_2 + c, 0) & \text{if } x_1 + x_2 + c < 10 \\ (x_2, x_1 + x_2 + c - 10, 1) & \text{if } x_1 + x_2 + c \geq 10 \end{cases}$$

For initial vectors $x = (x_1, x_2, 0)$ with $x_1 < x_2$ or $x = (x_1, x_2, 1)$ with $x_1 > x_2$ the sequence of iterates $x, f(x), f^2(x), \ldots$ is strictly periodic with period 108. If the initial vector x is not of those two types, and not $(0, 0, 0)$ or $(9, 9, 9)$, then the sequence beginning with $f(x)$ becomes periodic after a one iteration, with period 108, but the "seed" vector x may not reappear in the sequence. (Rules for assigning seed vectors so as to get strictly periodic sequences are in [6].)

The behavior of iterates of a function describing a random number generator can be illustrated by means of a directed graph. If the generator produces iterates $x, f(x), f^2(x), \ldots$ of a function f on elements x of a set X, then each element x of X is a node in the graph, connected by an arrow to $f(x)$.

We illustrate for the above two examples: the graph of the lagged-Fibonacci generator F(2,1,+ mod 10) follows in Figure 1, and the analogous add-with-carry generator, $x_n = x_{n-2} + x_{n-1} + c \bmod 10$, is graphed in Figure 2 on the opposite page.

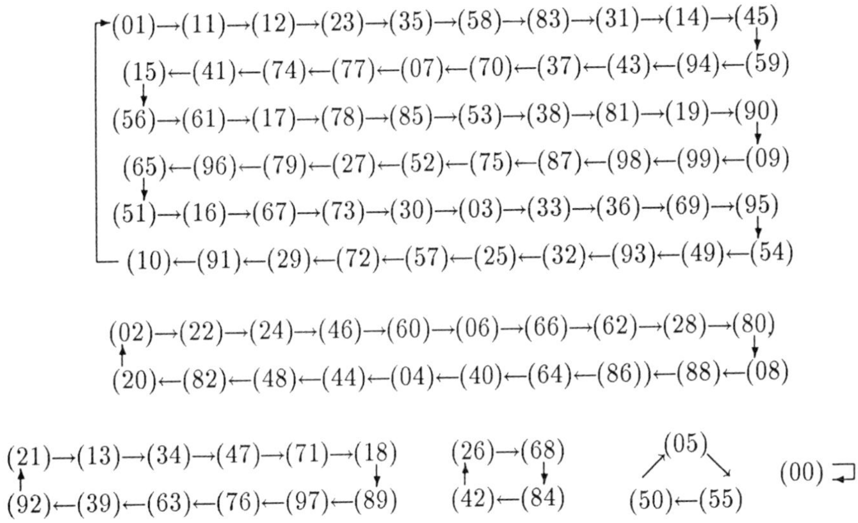

$$\begin{array}{l}
(01) \to (11) \to (12) \to (23) \to (35) \to (58) \to (83) \to (31) \to (14) \to (45) \\
(15) \leftarrow (41) \leftarrow (74) \leftarrow (77) \leftarrow (07) \leftarrow (70) \leftarrow (37) \leftarrow (43) \leftarrow (94) \leftarrow (59) \\
(56) \to (61) \to (17) \to (78) \to (85) \to (53) \to (38) \to (81) \to (19) \to (90) \\
(65) \leftarrow (96) \leftarrow (79) \leftarrow (27) \leftarrow (52) \leftarrow (75) \leftarrow (87) \leftarrow (98) \leftarrow (99) \leftarrow (09) \\
(51) \to (16) \to (67) \to (73) \to (30) \to (03) \to (33) \to (36) \to (69) \to (95) \\
(10) \leftarrow (91) \leftarrow (29) \leftarrow (72) \leftarrow (57) \leftarrow (25) \leftarrow (32) \leftarrow (93) \leftarrow (49) \leftarrow (54)
\end{array}$$

$$\begin{array}{l}
(02) \to (22) \to (24) \to (46) \to (60) \to (06) \to (66) \to (62) \to (28) \to (80) \\
(20) \leftarrow (82) \leftarrow (48) \leftarrow (44) \leftarrow (04) \leftarrow (40) \leftarrow (64) \leftarrow (86)) \leftarrow (88) \leftarrow (08)
\end{array}$$

$$\begin{array}{llll}
(21) \to (13) \to (34) \to (47) \to (71) \to (18) & (26) \to (68) & (05) & (00) \circlearrowleft \\
(92) \leftarrow (39) \leftarrow (63) \leftarrow (76) \leftarrow (97) \leftarrow (89) & (42) \leftarrow (84) & (50) \leftarrow (55) &
\end{array}$$

Figure 1.

The directed graph of the lagged-Fibonacci generator F(2,1,+ mod 10), that is, iterates of the function $f[(a,b)] = (b, a+b \bmod 10)$. The graph has six components (cycles) of lengths 60,20,12,4,3,1.

On the next page we get a completely different directed graph by changing the lagged-Fibonacci generator $x_n = x_{n-1} + x_{n-2} \bmod 10$ to the add-with-carry generator $x_n = x_{n-1} + x_{n-2} + c \bmod 10$. The iterating function f works on 3-tuples:

$$f(x_1, x_2, c) = \begin{cases} (x_2, x_1 + x_2 + c, 0) & \text{if } x_1 + x_2 + c < 10 \\ (x_2, x_1 + x_2 + c - 10, 1) & \text{if } x_1 + x_2 + c \geq 10 \end{cases}$$

The graph has three components, the trivial (000) and (991) together with the main component, which has 108 'seed' vectors that lead to a strictly periodic sequence, and 90 'seed' vectors that iterate to a cycle but have no antecedents. The generator has period 108 because 10 is a primitive root of $10^2 + 10 - 1$.

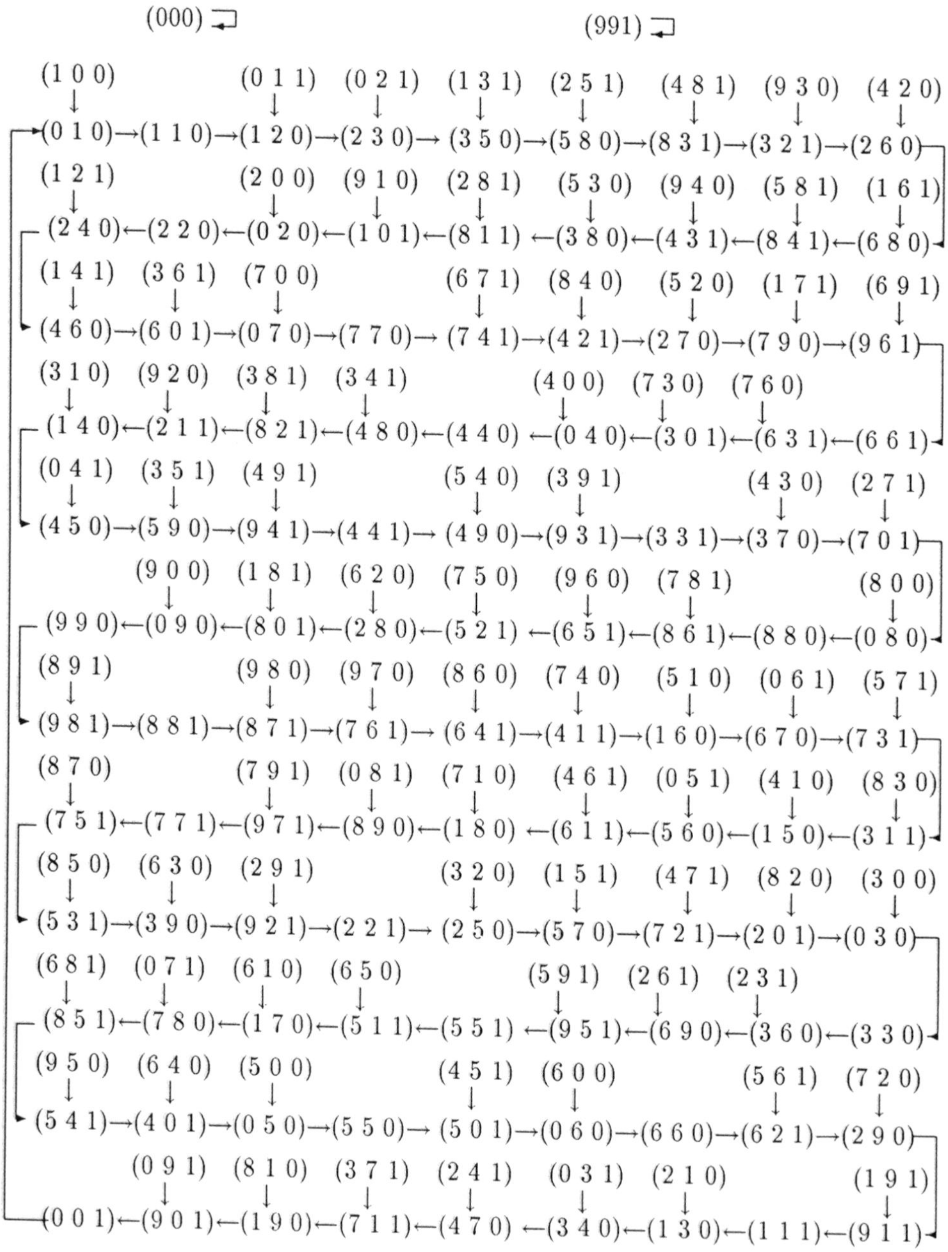

GEORGE MARSAGLIA

(000) ↵   (991) ↵

(1 0 0)   (0 1 1) (0 2 1) (1 3 1) (2 5 1)   (4 8 1) (9 3 0) (4 2 0)
   ↓   ↓   ↓   ↓   ↓   ↓   ↓   ↓
┌→(0 1 0)→(1 1 0)→(1 2 0)→(2 3 0)→ (3 5 0)→(5 8 0)→(8 3 1)→(3 2 1)→(2 6 0)┐
│ (1 2 1)   (2 0 0) (9 1 0) (2 8 1)   (5 3 0)   (9 4 0) (5 8 1) (1 6 1)│
│   ↓   ↓   ↓   ↓   ↓   ↓   ↓   ↓ │
┌ (2 4 0)←(2 2 0)←(0 2 0)←(1 0 1)←(8 1 1) ←(3 8 0)←(4 3 1)←(8 4 1)←(6 8 0)┘
│ (1 4 1) (3 6 1) (7 0 0)   (6 7 1) (8 4 0)   (5 2 0) (1 7 1) (6 9 1)
│   ↓   ↓   ↓   ↓   ↓   ↓   ↓   ↓
└ (4 6 0)→(6 0 1)→(0 7 0)→(7 7 0)→ (7 4 1)→(4 2 1)→(2 7 0)→(7 9 0)→(9 6 1)┐
  (3 1 0) (9 2 0) (3 8 1) (3 4 1)   (4 0 0) (7 3 0) (7 6 0)   │
     ↓   ↓   ↓   ↓   ↓   ↓   ↓   │
┌ (1 4 0)←(2 1 1)←(8 2 1)←(4 8 0)←(4 4 0) ←(0 4 0)←(3 0 1)←(6 3 1)←(6 6 1)┘
│ (0 4 1) (3 5 1) (4 9 1)   (5 4 0) (3 9 1)   (4 3 0) (2 7 1)
│   ↓   ↓   ↓   ↓   ↓   ↓   ↓
└ (4 5 0)→(5 9 0)→(9 4 1)→(4 4 1)→ (4 9 0)→(9 3 1)→(3 3 1)→(3 7 0)→(7 0 1)┐
  (9 0 0) (1 8 1) (6 2 0) (7 5 0) (9 6 0) (7 8 1)   (8 0 0)   │
     ↓   ↓   ↓   ↓   ↓   ↓   ↓   │
┌ (9 9 0)←(0 9 0)←(8 0 1)←(2 8 0)←(5 2 1) ←(6 5 1)←(8 6 1)←(8 8 0)←(0 8 0)┘
│ (8 9 1)   (9 8 0) (9 7 0) (8 6 0) (7 4 0)   (5 1 0) (0 6 1) (5 7 1)
│   ↓   ↓   ↓   ↓   ↓   ↓   ↓   ↓
└ (9 8 1)→(8 8 1)→(8 7 1)→(7 6 1)→ (6 4 1)→(4 1 1)→(1 6 0)→(6 7 0)→(7 3 1)┐
  (8 7 0)   (7 9 1) (0 8 1) (7 1 0) (4 6 1)   (0 5 1) (4 1 0) (8 3 0)│
     ↓   ↓   ↓   ↓   ↓   ↓   ↓   ↓ │
┌ (7 5 1)←(7 7 1)←(9 7 1)←(8 9 0)←(1 8 0) ←(6 1 1)←(5 6 0)←(1 5 0)←(3 1 1)┘
  (8 5 0) (6 3 0) (2 9 1)   (3 2 0) (1 5 1)   (4 7 1) (8 2 0) (3 0 0)
     ↓   ↓   ↓   ↓   ↓   ↓   ↓   ↓
└ (5 3 1)→(3 9 0)→(9 2 1)→(2 2 1)→ (2 5 0)→(5 7 0)→(7 2 1)→(2 0 1)→(0 3 0)┐
  (6 8 1) (0 7 1) (6 1 0) (6 5 0)   (5 9 1) (2 6 1) (2 3 1)   │
     ↓   ↓   ↓   ↓   ↓   ↓   ↓   │
┌ (8 5 1)←(7 8 0)←(1 7 0)←(5 1 1)←(5 5 1) ←(9 5 1)←(6 9 0)←(3 6 0)←(3 3 0)┘
  (9 5 0) (6 4 0) (5 0 0)   (4 5 1) (6 0 0)   (5 6 1) (7 2 0)
     ↓   ↓   ↓   ↓   ↓   ↓   ↓
└ (5 4 1)→(4 0 1)→(0 5 0)→(5 5 0)→ (5 0 1)→(0 6 0)→(6 6 0)→(6 2 1)→(2 9 0)┐
  (0 9 1) (8 1 0) (3 7 1) (2 4 1) (0 3 1) (2 1 0)   (1 9 1)   │
     ↓   ↓   ↓   ↓   ↓   ↓   ↓   │
└(0 0 1)←(9 0 1)←(1 9 0)←(7 1 1)←(4 7 0) ←(3 4 0)←(1 3 0)←(1 1 1)←(9 1 1)┘

Figure 2.
Directed graph of the add-with-carry generator
$$x_n = x_{n-1} + x_{n-2} + c \bmod 10$$

Can you discover the rule for vectors outside the main loop?

As with lagged-Fibonacci sequences, a whole class of add-with-carry generators can be created by altering the lags from the values $r = 2$ and $s = 1$ used in the previous example. The general add-with-carry generator has a base b, lags r and s with $r > s$, a seed vector $x = (x_1, x_2, \ldots, x_r, c)$ with elements 'digits' of the base b. Then the generated sequence is $x, f(x), f^2(x), f^3(x), \ldots$ with

$$f(x_1, \ldots, x_r, c) = \begin{cases} (x_2, \ldots, x_r, x_{r+1-s} + x_1 + c, 0) & \text{if } x_{r+1-s} + x_1 + c < b \\ (x_2, \ldots, x_r, x_{r+1-s} + x_1 + c - b, 1) & \text{if } x_{r+1-s} + x_1 + c \geq b. \end{cases}$$

With appropriately chosen base b, lags r and s and seed vector x, the generated sequence $x, f(x), f^2(x), \ldots$ will be periodic with period $b^r + b^s - 2$. These generators have extremely long periods. For example, when b is near 2^{32}, each base-b 'digit' is a computer word and with r around 20 or so then periods of some 2^{640} are attainable, at the cost of only r memory locations and simple computer arithmetic. (The add-with-carry instruction is basic to all CPU instruction sets.)

4.2   The New Class: subtract-with-borrow generators

The add-with-carry generators described above have periods $b^r + b^s - 2$, if $m = b^r + b^s - 1$ is prime and has b as a primitive root. Even if $m = b^r + b^s - 1$ is not prime, the period of the sequence will be the order of b in the group of residues relatively prime to m. Unfortunately, with m's of the sizes of interest, some 2^{640} or more, we have no available means to find the order of b unless we can factor $\phi(m)$. It is not difficult to find primes of the form $m = b^r + b^s - 1$, but factoring $\phi(m) = b^r + b^s - 2$ is not feasible when b is around 2^{32} and r from 20 to 50.

For that reason, (and for another reason that leads to direct floating point implementation of the new generators), we consider subtract-with-borrow generators of the form, with $r > s$,

$$x_n = x_{n-r} - x_{n-s} - c \bmod b \qquad \text{or} \qquad x_n = x_{n-s} - x_{n-r} - c \bmod b.$$

As we shall see, for proper choices of r and s, these lead to generators with periods $b^r - b^s - 2$ and $b^r - b^s$, respectively, and it is the latter case that holds the most promise.

So we concentrate on this kind of subtract-with-borrow generator: $x_n = x_{n-s} - x_{n-r} - c \bmod b$. If $m = b^r - b^s + 1$ is prime then that generator will have period the order of b in the group of non-zero residues of the prime m. We may be able to find the order of b because there is some hope of factoring $\phi(m) = m - 1 = b^r - b^s$.

5   Periods of the New Generators

The fundamental result for establishing periods comes from recognizing that these generators behave very much like the operation of long addition with carry (that some of us in the pre-calculator age learned in school). Once this addition is explicitly written, it is easy to recognize that the sequence of digits formed by the add-with-carry or subtract-with-borrow operation is, in reverse order, the same as the sequence of digits of the base-b expansion a proper fraction $k/(b^r \pm b^s \pm 1)$. Here r, s and b are the lags and the base, respectively, and choices of $\pm$ depend on the particular generator.

5.1 Results from number theory

We need some background material from number theory to establish periods of the new class of generators. This elementary material has been known for hundreds of years, but it is seldom mentioned in modern books. We summarize it here. It concerns the decimal expansions of fractions—expansions to a base b, rather than the customary base 10—but we illustrate with the more familiar base 10.

Let the modulus m be chosen and consider the group G of $\phi(m)$ reduced residues of m relatively prime to m. For k in G we want the base-b expansion of k/m. That expansion is strictly periodic with period the order of b in the group G. (This requires the assumption that $b \in G$.) The cyclic subgroup generated by b partitions G into cosets. Two elements g and h of G are equivalent (belong to the same coset) if $g = hb^j$ for some j. If two elements g, h belong to the same coset then g/m and h/m have the same base-b expansion with period the order of b, except that the 'digits' in their periods are cyclic permutations of one another.

Example: modulus $m = 39$, base $b = 10$. The powers of 10 mod m generate the cyclic subgroup $\{1, 10, 22, 25, 16, 4\}$, so the order of 10 for modulus 39 is 6. Successive elements k of that subgroup have a common set of digits in the period-6 decimal expansion of k/m, each a cyclic permutation of the previous one:

$1/39 = .025641025\ldots, \quad 10/39 = .256410256\ldots, \quad 22/39 = .564102564\ldots,$

$25/39 = .641025641\ldots, \quad 16/39 = .410256410\ldots, \quad 4/39 = .102564102\ldots.$

Now choose an element not in the first coset, say 2. Its coset is $\{2, 20, 5, 11, 32, 8\}$, and ratios k/m with k from that coset all have the same digits in their period-6 decimal expansions, shifted by one because of successive multiplications by the base 10:

$2/39 = .051282051\ldots, \quad 20/39 = .512820512\ldots, \quad 5/39 = .128205128\ldots,$

$11/39 = .282051282\ldots, \quad 16/39 = .820512820\ldots, \quad 8/39 = .205128205\ldots.$

For our purposes, we want to choose primes m for which b is a primitive root. Then the period is $m - 1$ for the base-b expansion of every proper fraction k/m.

5.2   Periods of the new generators: add-with-carry

We now prove that *the period of the add-with-carry sequence that produces 'digits' of the base b by means of the relation $x_n = x_{n-r}+x_{n-s}+c$ mod b is the period of the base-b expansion of k/m for some k in $1 \le k < m$ and $m = b^r + b^s - 1$.*

To fix ideas, consider the case $r = 2$, $s = 1$ for modulus $b = 10$: $x_n = x_{n-2} + x_{n-1} + c$ mod 10, initialized with the seed digits 1,2 and initial carry $c=0$. The sequence is: $1, 2, 3, 5, 8, 3, 2, 6, 8, 4, 3, 8, \ldots$. Now the idea behind the proof is that digits of the sequence that are $r - s$ positions apart are added, with carry, to form the next digit, moving left-to-right. We are all familiar with a similar arithmetic operation when adding two large integers, except that addition-with-carry moves right-to-left. To illustrate the analogy, let I be the integer whose digits are the first twelve of our sample sequence, *in reverse order*:

$$I = 834862385321.$$

Now shift I left $r - s = 1$ positions (form $10I$) and add it to I, getting

$$
\begin{array}{rcl}
I & = & 834862385321 \\
bI & = & 8348623853210 \\
\hline
I + bI & = & 9183486238531
\end{array}
$$

Because of the rule for forming the sequence, there will be a substring of digits common to each of the three levels; they are indicated in boldface. Let S be the integer formed by those digits—in this case, $S = 8348623853$. The appearance of S in each of the three levels of the sum enables us to develop a simple linear equation for S:

$$10^2 S + 21 + 10^3 S + 210 = 91 \times 10^{11} + 10^1 S + 1,$$

leading to
$$(10^2 + 10 - 1)S = 91 \times 10^{10} - 23.$$

Thus
$$S = 10^{10}\frac{91}{109} - \frac{23}{109}.$$

Now S is an integer, so the fractional part of $10^{10}(91/109)$ must cancel that of $23/109$, and it follows that S is the integer part of $10^{10}(91/109)$, that is, S's digits are the first ten digits of the decimal expansion of $91/109$.

$$\frac{91}{10^2 + 10^1 - 1} = .8348623853\ldots$$

We may apply such an argument to the reversed digits in an arbitrarily long finite string formed by $x_n = x_{n-r} + x_{n-s} + c$ mod b. The period of the sequence will be the period of the base-b expansion of a proper fraction of the form k/m with $m = b^r + b^s - 1$. The integer k will be that formed by the r digits on the left of the string S in the third row of the sum. When k is relatively prime to m the period will be the order of b for modulus m. Making m a prime ensures this, of course.

Here are some other examples. In the first two the sequence is

$$x_n = x_{n-4} + x_{n-2} + c \bmod 10, \qquad \text{seed } 7493 \rightarrow 74936852218305888372\ldots$$

with initial carry $c = 0$. Take, say, the first 20 digits of the sequence and write them in reverse order to get the integer I. Then form $b^{r-s}I$ by shifting I left two positions, bringing in two zeros. Then add. Because of the rule for forming new digits, the three levels will have a common string, indicated in bold face.

$$
\begin{aligned}
I &= \mathbf{27388850381225863}947 \\
b^2 I &= \mathbf{27388850381225863}94700 \\
\overline{I + b^2 I} &= 2766\mathbf{27388850381225}8647
\end{aligned}
$$

Let S be the integer formed by those digits. Presence of S in each of the three levels provides a linear equation:

$$b^4 S + 3947 + b^6 S + 394700 = 2766 b^{18} + b^2 S + 47,$$

leading to

$$(b^4 + b^2 - 1)S = 2766 b^{16} - 3908,$$

and thus

$$S = \frac{2766 b^{16}}{b^4 + b^2 - 1} - \frac{3908}{b^4 + b^2 - 1}.$$

Since S is an integer, the fractional part of the first term must cancel the second, and thus S is the first 16 digits of the expansion of $2766/(b^4 + b^2 - 1)$:

$$\frac{2766}{10^4 + 10^2 - 1} = .2738885038122586\ldots$$

Suppose, instead of 20, we take 15 digits of the above sequence and form an integer I by writing them in reverse order. We then add $b^{r-s}I$ to get this tableaux with the digits of S in boldface:

$$
\begin{aligned}
I &= \mathbf{850381225863}947 \\
b^2 I &= \mathbf{850381225863}94700 \\
\overline{I + b^2 I} &= 8\mathbf{5888503812258}647
\end{aligned}
$$

The linear equation in S that results from this sum is

$$b^4 S + 3947 + b^6 S + 394700 = 8588 b^{13} + b^2 S + 47,$$

from which, because S is an integer, we conclude that $S = 8588 b^{11}/(b^4 + b^2 - 1)$, that is, the digits of S are the first 11 digits in the expansion of $8588/10099$:

$$\frac{8588}{10^4 + 10^2 - 1} = .85038122586\ldots$$

Finally, here is one more tableaux with base $b = 6$ and lags $r = 6$ and $s = 3$:

$$x_n = x_{n-6} + x_{n-3} + c \bmod 6, \qquad \text{seed } 153024 \rightarrow 153024112230442 \ldots$$

$$
\begin{aligned}
I &= && 244032211420351 \\
b^3 I &= && 244032211420351000 \\
\overline{I + b^3 I} &= && 244320244032211351
\end{aligned}
$$

$$\frac{244320}{6^4 + 6^3 - 1} = .244032211 \ldots \text{ base } 6$$

5.3 Periods of the generators: subtract-with-borrow

For sequences generated by $x_n = x_{n-r} - x_{n-s} - c \bmod b$, the period is also the period of a proper fraction k/m, where this time $m = b^r - b^s - 1$. This may be established in a manner similar to that for add-with-carry sequences: Choose r starting values and generate a sequence of arbitrary length. As before, let I be the integer formed by putting those digits in reverse order. For example, with $b = 10$, $r = 5$, $s = 3$ and starting values $5, 9, 7, 7, 7$ the reversed string of 15 digits yields $I = 304285901877795$. Shift I left by s positions (multiply by b^s) and add to get

$$
\begin{aligned}
I &= && 304285901877795 \\
b^s I &= && 304285901877795000 \\
\overline{I + b^s I} &= && 304590187779672795
\end{aligned}
$$

The boldface string $S = 59018779$ appears in each line. Cancel the leading 304 and the trailing 5 to get a simplified equation for S:

$$304285 b^9 + S + 28 b^{12} + b^3 S + 500 = b^5 S + 67279.$$

Thus

$$584285 b^9 - 64779 = (b^5 - b^3 - 1)S,$$

and, with $m = b^5 - b^3 - 1 = 98999$,

$$S = 10^9 \frac{58428}{98999} - \frac{66779}{98999}.$$

Since S is an integer, it must be the integer part of the first term, that is, the digits of S are the leading 9 digits of the decimal expansion of $58428/98999$: $S = 590187779$.

We may apply such an argument to the reversed string of an arbitrary length sequence with r initial values and, for $n > r$, $x_n = x_{n-r} - x_{n-s} - c \bmod b$. Its period will be the period of the base-b expansion of a proper fraction k/m, with $m = b^r - b^s - 1$. (The particular value of k changes with the length of the string used to form I. It is of no particular importance—the actual value is that of the trailing 5 digits of $J + b^s J$, where J is the integer formed by the leading r

digits of I. The important point is that the linear equation always has S with coefficient $b^d m$, for some d and $m = b^r - b^s - 1$, so that the solution for the integer S is the first d digits of a proper fraction k/m for some k.)

For the case $x_n = x_{n-s} - x_{n-r} - c$ with $r > s$ a similar derivation holds, except that I is added to $b^r I$. The common string S has a solution of the form $S = b^d \frac{k}{m} + \theta$ with $m = b^r - b^s + 1$ and $0 < \theta < 1$. Then the base-b digits of S are again obtained from the leading digits of the base-b expansion of a proper fraction k/m. We illustrate with a final tableaux:

$$x_n = x_{n-3} - x_{n-5} - c \bmod 10, \qquad \text{seed } 26479 \rightarrow 264792155124662426791534\ldots$$

$$
\begin{aligned}
I &= & 97624266421551297462 \\
b^r I &= & 97624266421551297462\underline{00000} \\
\hline
I + b^r I &= & 97625242664215512974\,97462
\end{aligned}
$$

$$(b^5 - b^3 + 1)S = 24024 b^{15} - 6102538$$

$$\frac{24024}{10^5 - 10^2 + 1} = .242664215512974\ldots$$

The rule for forming k may be deduced by noting that $24024 = 25000 - 976$.

5.4 Summary for the new generators

The add-with-carry generator is described informally as $x_n = x_{n-r} + x_{n-s} + c \bmod b$. It is initialized with r seed values $x_1, \ldots, x_r$, not all 0 mod b, and an initial carry $c \in \{0, 1\}$. The rule for forming x_{n+1} and the new carry c is:

> Form $t = x_{n-r} + x_{n-s} + c \bmod b$.
> If $t < b$ then put $x_{n+1} = t$ and $c = 0$,
> Else put $x_{n+1} = t - b$ and $c = 1$.

The period of this generator is the order of b in the group of residues relatively prime to $m = b^r + b^s - 1$.

The most useful subtract-with-borrow generator is described informally as $x_n = x_{n-s} - x_{n-r} - c \bmod b$ with $r > s$. It is initialized with r seed values $x_1, \ldots, x_r$, not all 0 mod b, and an initial borrow $c \in \{0, 1\}$. The rule for forming x_{n+1} and the new borrow c is:

> Form $t = x_{n-s} - x_{n-r} - c \bmod b$.
> If $t \geq 0$ then put $x_{n+1} = t$ and $c = 0$,
> Else put $x_{n+1} = t + b$ and $c = 1$.

The period of this generator is the order of b in the group of residues relatively prime to $m = b^r - b^s + 1$.

The other subtract-with-borrow generator is, informally, $x_{n+1} = x_{n-r} - x_{n-s} - c \bmod b$. Each x_{n+1} is generated in a matter analagous to the $x_{n-s} - x_{n-r} - c \bmod b$ generator, and the period is the order of b in the group of residues prime to $m = b^r - b^s - 1$. It is less promising because of the difficulty of factoring $m - 1$ for prime m.

The table below lists examples of some of the most succesful kinds of random number generators. Lines 1-3 are examples of three of the most frequently used congruential generators; line 4 is an example of an extended congruential generator with prime modulus p and period $p^2 - 1$. Lines 6-12 give examples of lagged-Fibonacci generators with increasingly long periods, except for line 8, which shows the drastic reduction of period arising from use of the exclusive-or operation rather than subtraction or multiplication.

The tremendously long periods of subtract-with-borrow generators are exemplified in lines 13-17, and line 18 gives an example of a very good generator that arises from combining two simple generators that individually are not very promising.

Examples of various kinds of random number generators

	Seeds: Number and Type	Type and Generating Rule	Period
		Congruential	
1	1 32-bit odd integer	$x_n = 69069x_{n-1} \bmod 2^{32}$	2.1×10^9
2	1 32-bit integer	$x_n = 69069x_{n-1} + 1 \bmod 2^{32}$	4.3×10^9
3	1 31-bit integer$\neq 0$	$x_n = 16807x_{n-1} \bmod 2^{31} - 1$	2.1×10^9
		Extended Congruential	
4	2 31-bit integers	$x_n = 1999x_{n-1} + 4444x_{n-2} \bmod 2^{31} - 1$	4.6×10^{18}
		Lagged Fibonacci	
6	17 32-bit integers	$x_n = x_{n-17} - x_{n-5} \bmod 2^{32}$	2.8×10^{14}
7	17 32-bit odd integers	$x_n = x_{n-17} * x_{n-5} \bmod 2^{32}$	7.0×10^{13}
8	17 32-bit integers	$x_n = x_{n-17} \oplus x_{n-5} \bmod 2^{32}$	1.3×10^5
9	55 32-bit integers	$x_n = x_{n-55} - x_{n-24} \bmod 2^{32}$	7.7×10^{25}
10	55 32-bit odd integers	$x_n = x_{n-55} * x_{n-24} \bmod 2^{32}$	1.9×10^{25}
11	97 reals	$x_n = x_{n-97} - x_{n-33} \bmod 1$	1.3×10^{36}
12	607 32-bit integers	$x_n = x_{n-607} - x_{n-273} \bmod 2^{32}$	10^{192}
		Subtract-with-Borrow	
13	847 bits	$x_n = x_{240} - x_{847} - c \bmod 2$	9.4×10^{254}
14	1751 bits	$x_n = x_{n-472} - x_{n-1751} - c \bmod 2$	1.3×10^{527}
15	37 32-bit integers	$x_n = x_{n-24} - x_{n-37} - c \bmod 2^{32}$	4.1×10^{354}
16	30 64-bit integers	$x_n = x_{n-6} - x_{n-30} - c \bmod 2^{64}$	9.5×10^{577}
17	39 reals	$x_n = x_{n-25} - x_{n-39} - c \bmod 1$	8.6×10^{278}
		Combination	
18	2 32-bit x's, odd 3 y's$< 2^{30} - 35$	$x_n = x_{n-1} * x_{n-2} \bmod 2^{32}$ $y_n = y_{n-3} - y_{n-1} \bmod 2^{30} - 35$ $z_n = x_n - y_n \bmod 2^{32}$	2.3×10^{18}

REFERENCES

[1] Knuth, Donald E., (1981), *The Art of Computer Programming: Volume 2: Seminumerical Algorithms*, Second edition, Reading: Addison Wesley, 688p.

[2] George Marsaglia, Random numbers fall mainly in the planes, (1968), *Proceedings of the National Academy of Sciences*, **61**, 25–28.

[3] Marsaglia, George, (1972), The structure of linear congruential sequences, *Applications of Number Theory to Numerical Analysis*, Z. K. Zaremba, ed., New York: Academic Press, 1972, 489p.

[4] Marsaglia, George and Tsay, L. H., (1985), Matrices and the structure of random number sequences *Linear Algebra and its Applications*, **67**, 147–156.

[5] Marsaglia, George, (1985), A Current View of Random Number Generators, Keynote Address: *Proceedings, Computer Science and Statistics: 16th Symposium on the Interface*, New York: Elsevier.

[6] Marsaglia, George and Zaman, Arif, (1991), A new class of random number generators, *Annals of Applied Probability*, **1** No. 3, 462-480.

[7] Ripley, Brian D., (1987), *Stochastic Simulation*, New York: J. Wiley, 237p.

Proceedings of Symposia in Applied Mathematics
Volume 46, 1992

Cyclotomy and Cyclic Codes

Vera Pless

1. Introduction. The history of error-correcting codes began in 1948 with the publication of a famous paper by Claude Shannon. Shannon demonstrated that for certain channels one could communicate as reliably as one wanted by using long enough error-correcting codes. Shannon's achievement was in showing that very good codes must exist not in constructing them. The first efforts in coding theory showed that it was indeed quite difficult to find these codes. Even though the primary motivation was practical, algebraic tools were introduced early in the fifties and sixties at which time cyclic codes and many good specific codes were discovered. Also many properties of these codes were developed. At present error correcting codes are widely used in communication systems and large digital systems and they are studied by engineers interested in practical applications. At the same time a mathematical theory of error-correcting codes has been developed by mathematicians who became interested in these fascinating new objects with important relations to combinatorial designs, finite groups, lattices, sphere packings and many other branches of mathematics. Number theory has entered into coding theory in many ways, too many to be covered by this short article. We concentrate on demonstrating how new properties of cyclic codes depend on number theoretic results. We will mention other relations of codes to number theory more briefly.

2. Basic concepts. An (n, k) *error-correcting code over* $GF(q)$ is a k-dimensional subspace of the space of n-tuples whose components are elements of $GF(q)$. For practical use and in theoretical investigations $GF(2)$ plays a special role so that many of our examples and results will be of *binary* codes. As a code is a vector space it can be described by a basis which in coding terms is called a *generator matrix*.

An important concept in coding is the *weight* of a vector, that is the number of non-zero components it has. The weight of x is denoted by $wt(x)$. The smallest non-zero weight of a vector in a code is called the *minimum weight* of the code and is often denoted by d. When the minimum weight d is known we say we have

1991 *Mathematics Subject Classification.* Primary 94B15; Secondary 11B75.

This paper is in final form and no version of it will be submitted for publication elsewhere.

Supported in part by Grant NSA-MDA 904-91-H-0003.

© 1992 American Mathematical Society
0160-7634/92 $1.00 + $.25 per page

an (n, k, d) code. For example the following is a generator matrix for a binary $(7, 4, 3)$ code called the Hamming code.

$$
G = \begin{array}{c}
\begin{array}{ccccccc} 0 & 1 & 2 & 3 & 4 & 5 & 6 \end{array} \\
\left(\begin{array}{ccccccc}
0 & 1 & 1 & 0 & 1 & 0 & 0 \\
0 & 0 & 1 & 1 & 0 & 1 & 0 \\
0 & 0 & 0 & 1 & 1 & 0 & 1 \\
1 & 0 & 0 & 0 & 1 & 1 & 0
\end{array} \right)
\end{array}
$$

The reason the minimum weight is so important is that it is known [12] that a code with minimum weight d can correct $\lfloor \frac{d-1}{2} \rfloor$ errors. Hence a measure of the "goodness" of a code is how large d is, i.e., given n and k we want d as large as possible. Also in constructing a code or a family of codes it is desirable to have some knowledge about the possible weights of vectors in the code.

If C is an (n, k) code, $C^{\perp}$ is the $(n, n-k)$ code consisting of all vectors which are orthogonal to the vectors in C with respect to the usual inner product. A basis of $C^{\perp}$ is called a *parity-check* matrix for C and it arises naturally in the practical situation of determining the vector sent from the vector received (decoding). Clearly the vectors in C can be described as being all vectors orthogonal to the rows of a parity-check matrix for C. For example, the following is a parity-check matrix for the $(7, 4, 3)$ Hamming code described above. It is a basis of the $(7, 3, 4)$ code orthogonal to this code.

$$
H = \begin{array}{c}
\begin{array}{ccccccc} 0 & 1 & 2 & 3 & 4 & 5 & 6 \end{array} \\
\left(\begin{array}{ccccccc}
1 & 0 & 0 & 1 & 0 & 1 & 1 \\
1 & 1 & 0 & 0 & 1 & 0 & 1 \\
1 & 1 & 1 & 0 & 0 & 1 & 0
\end{array} \right)
\end{array}
$$

A code C is called *self-orthogonal* if $C \subset C^{\perp}$. Clearly all vectors in a binary self-orthogonal code have even weight. The $(7, 3, 4)$ code with basis H above can be easily checked to be self-orthogonal. A code C is called *self-dual* if $C = C^{\perp}$. A self-dual code is an $(n, n/2)$ code and so can only exist for even n.

In order to give an example of a binary self-dual code we define the *extended code* of a binary (n, k) code C, $\widehat{C}$ to be the $(n+1, k)$ binary code with an overall parity check added. This is a 0 if the codeword has even weight and a 1 if it has odd weight. We see that all vectors in $\widehat{C}$ have even weight.

The extended Hamming $(7, 4, 3)$ code is an $(8, 4, 4)$ code with the following generator matrix.

$$
\widehat{G} = \left(\begin{array}{cccccccc}
1 & 0 & 1 & 1 & 0 & 1 & 0 & 0 \\
1 & 0 & 0 & 1 & 1 & 0 & 1 & 0 \\
1 & 0 & 0 & 0 & 1 & 1 & 0 & 1 \\
1 & 1 & 0 & 0 & 0 & 1 & 1 & 0
\end{array} \right).
$$

Notice that all rows of $\widehat{G}$ have weight 4 and every two rows have an even number of ones in common. This means that the extended Hamming code is self-orthogonal and hence by its dimension self-dual. It is not difficult to see that there is one vector of weight 0 and one vector of weight 8 in this code and that the other 14 vectors have weight 4. Hence the weights of all non-zero vectors are divisible by 4.

A binary code in which all vectors have weight divisible by 4 is called *doubly even*. It can easily be shown that a doubly-even code must be self-orthogonal.

Self-dual codes are mathematically interesting, many of the best algebraic codes are self-dual. Also, self-dual codes or codes whose extensions are self-dual have many connections to interesting combinatorial designs and finite groups. These are some of the reasons why these codes have been widely studied and classified [2, 12]. There are relations, not yet completely understood, between doubly-even, self-dual codes and even unimodular lattices. These lattices have been classified up to length 24 [2] and there are similarities between these classifications and that of the doubly-even, self-dual codes. The method of classifying is similar and some of the lattices which occur are generated by the doubly-even codes. Further, the theta function of these lattices has similarities to the weight distribution of these codes.

The set of all permutations of the coordinate indices which send the vectors of a code onto other vectors in the code forms a group called the *group* of the code. The coordinate permutations are the linear transformations which preserve weight, an important invariant for a code. It is easy to see that a code C and its orthogonal code have the same group. Thus the $(7, 4, 3)$ Hamming code with generator matrix G and the $(7, 3, 4)$ code with generator matrix H have the same group which has order 168. The extended Hamming $(8, 4, 4)$ code with generator matrix $\hat{G}$ has a group of order 1,344. This group is known to be doubly-transitive on the coordinate indices and the stabilizer of a point has order 168.

The rows of G are cyclic shifts of the first row so one might think that this $(7, 4, 3)$ code has the coordinate permutation (the cyclic shift) $i \to i + 1 \pmod 7$ in its group. This can easily be checked to be so as it is only necessary to check this on the rows of G (the basis of the code). The check for the first 3 rows is obvious, it can also be shown that the cyclic shift of the last row is a combination of the rows of G, hence in the code. By linearity then the cyclic shift is in the group of this code.

One definition of a cyclic code is the following. Let C be a length n code and label the coordinates $0, 1, \ldots, n - 1$. Then C is a *cyclic code* if the coordinate permution $i \to i + 1 \pmod n$ is in the group of C. This means that the entire cyclic group of order n in its natural representation on the coordinates is in the group of C.

If C is a cyclic code, so is $C^\perp$. The $(7, 4, 3)$ and $(7, 3, 4)$ codes with generator matrices G and H are cyclic codes. We note that 7 divides 168. Clearly other coordinate permutations are in the group of these codes.

3. Cyclic Codes. The consequences of this definition are quite striking. As before we label the coordinate indices $0, 1, \ldots, n - 1$. We can rephrase our previous definition of a cyclic code as follows. C is a *cyclic code* if $(c_0, c_1, \ldots, c_{n-1})$ is in C implies that $(c_{n-1}, c_0, \ldots, c_{n-2})$ is also in C. As we saw, our examples of a $(7, 4, 3)$ code and its orthogonal $(7, 3, 4)$ code are cyclic. The cycle form of the cyclic shift on 7 coordinate positions is $(0,1,2,3,4,5,6)$.

The following algebraic connection was discovered early on in coding theory. This was an exciting discovery with important practical implications. Let F denote $GF(q)$ and $F[x]$ denote the set of all polynomials in x with coefficients in F. We restrict ourselves to the situation where g. c. d. $(q, n) = 1$ and n is odd. Then $R_n = F[x]/(x^n - 1)$ is the set of all polynomials in x of degree $< n$ with coefficients in F. It is elementary to show that R_n, with the usual definition of polynomial addition and multiplication mod $(x^n - 1)$, is a principal ideal ring (P.I.R.) [12]. Associate a vector $(c_0, c_1, \dots, c_{n-1})$ in a cyclic code over $GF(q)$ with the polynomial $c_0 + c_1 x + \dots + c_n x^{n-1}$ in R_n. The key observation, which can be shown easily, is that a cyclic code corresponds to an ideal in R_n. For the case $n = 7$, $q = 2$ we associate the vector $(0, 1, 1, 0, 1, 0, 0)$ to the polynomial $x + x^2 + x^4$. Note that the cyclic shift of this vector, $(0, 0, 1, 1, 0, 1, 0)$ is associated to the polynomial $x^2 + x^3 + x^5 = (x + x^2 + x^4)x$. The fact that a cyclic shift corresponds to multiplying by x is the basis of this key observation.

Since R_n is a P.I.R., every vector in a cyclic code is a multiple of a generator polynomial. A P.I.R. can have more than one generator. Two of these are distinguished. One of them can be shown to be a factor of $x^n - 1$, that is every monic factor of $x^n - 1$ is a generator polynomial of a distinct cyclic code. This has the important practical consequence that in order to find generators of all cyclic codes of length n one has to factor $x^n - 1$. These factors have the important added advantage that they give the dimensions of the cyclic codes they generate. It is not pleasant to factor $x^n - 1$ over $GF(q)$ by hand though it can be done by computer. It has to be done for each n in turn. The other distinguihed generator, the idempotent generator, seems to have a more algebraic character. The following fact is well known.

Fact 1. R_n is a direct sum of minimal ideals. Hence any ideal, i.e. cyclic code can be expressed uniquely as a sum of minimal ideals.

It is easy to show that any cyclic code has an *idempotent generator* [12], that is a generating polynomial $e(x)$ which is also an idempotent, i.e. $e(x)^2 = e(x)$. The idempotent generator is actually the multiplicative unit of its cyclic code. If C has $e(x)$ as its idempotent generator we write $C = \langle e(x) \rangle$. The (7,4,3) binary Hamming code in our example has idempotent generator $e(x) = x + x^2 + x^4$. Note that $(x + x^2 + x^4)^2 = x^2 + x^4 + x = e(x)$. The (7,3,4) orthogonal code has idempotent generator $(1 + x^3 + x^5 + x^6)$. The following fact is easy to verify.

Fact 2. If C_1 and C_2 are cyclic codes with $C_1 = \langle e_1 \rangle$ and $C_2 = \langle e_2 \rangle$, then $C_1 \cap C_2 = \langle e_1 e_2 \rangle$ and $C_1 + C_2 = \langle e_1 + e_2 - e_1 e_2 \rangle$.

Note that if C_1 and C_2 are miniml ideals, $C_1 \cap C_2 = 0$ and $C_1 + C_2 = \langle e_1 + e_2 \rangle$.

Much information about cyclic codes is contained in cyclotomic cosets. Given n and q, the *q-cyclotomic cosets for n* are the disjoint sets

$$(0), (1, q, q^2, \dots), \dots, (r, qr, q^2 r, \dots)$$

where the elements in each set are computed (mod n). For example the 2-cyclotomic cosets for 7 are

$$(0), (1, 2, 4), (3, 6, 5),$$

the 3-cyclotomic cosets for 13 are

$$(0), (1, 3, 9), (2, 6, 5), (4, 12, 10), (7, 8, 11).$$

We just say cyclotomic cosets when the q and n are clear.

Fact 3. The number of irreducible factors of $(x^n - 1)$ over $GF(q)$ is the number, r, of q-cyclotomic cosets for n. Hence the number of cyclic codes of length n over $GF(q)$ is r^q. Further the dimensions of these codes can be computed from the sizes of the cyclotomic cosets.

Fact 4. If p is a prime, then all the q-cyclotomic cosets for p have the same size s except for (0) which has size 1. Further $rs = (p - 1)$.

In this case all the minimal ideals, i.e. minimal cyclic codes have dimension s except for one of dimension 1.

Thus there are 8 binary cyclic codes of length 7, 1 of dimension 0, 1 of dimension 1, 2 of dimension 3, 2 of dimension 4, 1 of dimension 6 and the whole space of dimension 7.

Let $h = (1, \ldots, 1)$ be the all one vector. Then $\langle 1/nh \rangle$ is always a cyclic code of dimension 1 and length n. Note that $1/n$ is computed in $GF(q)$.

If a is such that g. c. d. $(a, n) = 1$, then

$$\mu_a : i \to ai (\mathrm{mod}\ n)$$

is a coordinate permutation. We call such coordinate permutations *multipliers*. The study of multipliers in relation to cyclic codes seems to be new [12, 13] although they have been studied in the context of difference sets [5].

Fact 5. If g. c. d. $(a, n) = 1$, μ_a sends any cyclic code to another cyclic code.

If C is a cyclic code with idempotent generator $e(x)$, then $\mu_a(C)$ has idempotent generator $\mu_a(e(x))$. So μ_a is in the group of C iff $\mu_a(e(x)) = e(x)$.

Fact 6. If C is a cyclic code over $GF(q)$ then μ_q is in the group of C.

Fact 7. If C is a cyclic code with idempotent $e(x)$, $C^\perp = \langle 1 - \mu_{-1}(e(x)) \rangle$.

For the case $n = 7$, $q = 2$, $e(x) = x + x^2 + x^4$, we have $1 - \mu_{-1}(e(x)) = 1 + x^3 + x^5 + x^6$. The first idempotent $e(x)$ is the idempotent of the $(7,4,3)$ Hamming code and the second idempotent is that of its orthogonal code. Also μ_2 is in the group of each code. In cycle form μ_2 is $(1,2,4)$ $(3,6,5)$.

We note that constructing binary idempotents from cyclotomic cosets is extremely easy. Any idempotent is of the form $\sum x^i$ where the i are in a union of cyclotomic cosets. It is easy to verify that a polynomial of this form must indeed be an idempotent since if i occurs as a power so does $2i$. Using this observation we can see how easy it is to compute the idempotents of the 8 binary codes of length 7. In general, however, we do not have much information about the codes generated. Only in special situations do we know the dimension.

The dimension was known for quadratic residue codes which can be defined by their idempotents. For simplicity we define *binary quadratic residue codes*. Quadratic residues codes over an arbitrary finite field will be discussed in the next section. These are only defined of prime length p. When the quadratic residues $\mathbf{Q}$ in $GF(p)$ are a union of cyclotomic cosets then so are the non-residues $\mathbf{N}$. Let $e = \sum x^i$ where $i \in Q$ and $e' = \sum x^i$ where $i \in N$. Then there are 4 quadratic residue codes of length p. These are $\langle e \rangle$, $\langle 1 + e \rangle$, $\langle e' \rangle$ and $\langle 1 + e' \rangle$. We can give their dimensions in general; two have dimension $\frac{p+1}{2}$ and two have dimension $\frac{p-1}{2}$. Clearly they can only exist when the quadratic residues are a union of cyclotomic cosets which occurs when 2 is a quadratic residue (mod p). This is known to be the case for $p \equiv \pm 1 \pmod 8$. So binary quadratic residue codes exist of length p when $p \equiv \pm 1 \pmod 8$ and the description of the codes of dimension $\frac{p-1}{2}$ is different for $p \equiv 1 \pmod 8$ from that for $p \equiv -1 \pmod 8$. The (7,4,3) and (7,3,4) binary Hamming codes we have seen are quadratic residue codes as 1, 2, and 4 are the quadratic residues in $GF(7)$.

4. Duadic codes and generalizations.

Duadic codes are a generalization of quadratic residue codes. We call a vector $v = (v_0, \ldots, v_{n-1})$ *even-like* if

$$\sum_{i=0}^{n-1} v_i = 0 \text{ in } GF(q).$$

Otherwise it is called *odd-like*. If $q = 2$, an even-like vector has even weight and an odd-like vector has odd weight. We call a *cyclic code* *odd-like* if it contains odd-like vectors, otherwise we say it is even-like.

Suppose $C_1 = \langle e_1 \rangle$ and $C_2 = \langle e_2 \rangle$ are odd-like cyclic codes of length n with the following two properties.

1) There is an a with g. c. d. $(a, n) = 1$, $\mu_a(C_1) = C_2$ and $\mu_a(C_2) = C_1$.
2) $e_1 + e_2 = 1 + 1/n\ h$.

Then C_1 and C_2 are *odd-like duadic codes*. Also $C_1' = \langle 1 - e_e \rangle$ and $C_2' = \langle 1 - e_1 \rangle$ are *even-like duadic codes*. [13].

Clearly $\mu_a(C_1') = C_2'$, $\mu_a(C_2') = C_1'$ and if $e_1' = 1 - e_2$ and $e_2' = 1 - e_1$, then $e_1' + e_2' = 1 - 1/n\ h$.

It is not difficult to demonstrate the following theorem.

THEOREM 1. *If C_1 and C_2 are odd-like duadic codes of length n, then $\dim C_1 = \dim C_2 = \frac{n+1}{2}$. If C_1' and C_2' are even-like duadic codes, then $\dim C_1' = \dim C_2' = \frac{n-1}{2}$.*

PROOF We indicate how easy this is to prove. Multiply equation 2) by e_1. This gives $e_1 e_2 = 1/n\ e_1 h$. Since e_1 is an odd-like vector, it is easy to show that $e_1 h = h$ so that $e_1 e_2 = 1/n\ h$. Hence

$$\dim C_1 \cap C_2 = \dim \langle e_1 e_2 \rangle = 1.$$

Now

$$\dim(C_1 + C_2) = \dim \langle e_1 + e_2 - e_1 e_2 \rangle = \dim V$$

as $e_1 + e_2 - e_1 e_2 = 1$. We have $\dim C_1 = \dim C_2$ since $C_2 = \mu_a(C_1)$. Hence

$$\dim C_1 = \dim C_2 = \frac{n+1}{2}.$$

A similar argument holds for C_1' and C_2' after noting that $e_1' e_2' = 0$.

It is equally easy to show that $C_i' \subseteq C_i$ and that $C_i = C_i' + \langle 1/n\ h \rangle$.   Q.E.D.

Let q, n and a be as above and suppose that S_1 and S_2 are unions of cyclotomic cosets with the following two properties.

1) $S_1 \cap S_2 = \emptyset$, $S_1 \cup S_2 = \{1, 2, \ldots, n-1\}$.

2) $\mu_a(S_1) = S_2$ and $\mu_a(S_2) = S_1$.

Then S_1 and S_2 are called a *splitting* given by μ_a.

The next theorem gives the connection to duadic codes.

THEOREM 2 [10]. . *Duadic codes exist if and only if a splitting exists.*

Thus we can see by just examining the cyclotomic cosets whether splittings, and hence duadic codes, exist. For $GF(2)$, we can actually construct the idempotents of duadic codes directly from a splitting. The four idempotents are
$$\epsilon + \sum_{j \in S_i} x^j, \quad i = 1, 2, \quad \epsilon = 0, 1.$$
For example, if $p = 31$, the cyclotomic cosets are

$$C_1 = (1, 2, 4, 8, 16), \qquad\qquad C_3 = (3, 6, 12, 24, 17),$$
$$C_5 = (5, 10, 20, 9, 18), \qquad\qquad C_7 = (7, 14, 28, 25, 19),$$
$$C_{11} = (11, 22, 13, 26, 21), \qquad\qquad C_{15} = (15, 30, 29, 27, 23).$$
$$r = 6, s = 5$$

There are 6 splittings   given by

$$C_1 \cup C_5 \cup C_7 \text{ and } C_3 \cup C_{15} \cup C_{11} \qquad \mu_3 \text{ or } \mu_{-1}$$
(These give the quadratic residue codes)
$$C_1 \cup C_3 \cup C_5 \text{ and } C_{15} \cup C_7 \cup C_{11} \qquad \mu_{30} = \mu_{-1}$$
$$C_3 \cup C_5 \cup C_{10} \text{ and } C_7 \cup C_{11} \cup C_1 \qquad \mu_{-1}$$
$$C_5 \cup C_{15} \cup C_7 \text{ and } C_{11} \cup C_1 \cup C_3 \qquad \mu_{-1}$$
$$\text{two more} \qquad\qquad \mu_{-1}$$

Their 12 idempotents can be constructed as described and the 6 cyclic codes with the odd weight idempotents have dimension 16 while the 6 cyclic codes with the even weight idempotents have dimension 15.

If $q = 3$ and the length is $p = 13$, we can easily compute the cyclotomic cosets:

$$(1, 3, 9), (2, 6, 5), (4, 12, 10), (7, 8, 11); \quad r = 4, s = 3$$

μ_{-1} again gives a splitting so we know that duadic codes of dimensions 6 and 7 exist, but there is no easy way to construct their idempotents.

THEOREM 3. *Duadic codes of length n exist over $GF(q)$ if and only if q is a square($\bmod$ n).*

PROOF We just give an indication of the proof when n is a prime p. As before let r be the number of cyclotomic cosets. If q is a square mod p, then both the quadratic residues and nonresidues are unions of cyclotomic cosets so that r must be even.

Suppose that r is even, and let G denote the cyclic group of nonzero elements in $GF(p)$. Let H denote the subgroup generated by q. Then r is the index of H in G. As G is cyclic any subgroup of even index is contained in the subgroup of quadratic residues.

Hence we have shown that q is a square mod p iff r is even. It is not difficult to show that a splitting exists iff r is even.

Q.E.D.

Duadic codes were generalized to triadic codes [14] if there are three even-like codes $C_i = \langle e_i \rangle$, $i = 1, 2, 3$ and a multiplier μ_b so that $\mu_b(C_1) = C_2$, $\mu_b(C_2) = C_3$, $\mu_b(C_3) = C_1$ and $e_1 + e_2 + e_3 - 2e_1e_2e_3 = 1 - \frac{1}{n}h$. Again the dimensions of these codes can be determined and it was shown that triadic codes of prime length p exist iff q is a cubic residue mod p. This was further generalized to m-adic residue codes [1].

As remarked before self-dual codes are an important class of codes many of which have relations to conbinatorial designs and even unimodular lattices. So it is interesting to be able to determine when extended cyclic, self-dual codes exist and, in certain cases, to be able to construct their idempotents.

THEOREM 4 [10]. C' is a self-orthogonal, cyclic $\left(n, \frac{n-1}{2}\right)$ code if and only if C' and $(C')^{\perp} = C$ are even-like and odd-like duadic codes with splitting given by μ_{-1}.

PROOF We give a short indication of the easy proof. Suppose $C' = \langle 1 - e \rangle$ is a self-orthogonal $\left(n, \frac{n-1}{2}\right)$ code. Then $C = (C')^{\perp} = \langle \mu_{-1}(e) \rangle$. It is easy to show that C' is even-like and h must be orthogonal to C', hence in C. By the obvious dimension arguments $C = C' + \langle \frac{1}{n}h \rangle$. Thus the idempotent of C is $1 - e + \frac{1}{n}h$ which equals $\mu_{-1}(e)$. This shows that $C = \langle \mu_{-1}(e) \rangle$ and $C_2 = \langle e \rangle$ are duadic codes with splitting given by μ_{-1}. The even-like duadic codes for this splitting are $C' = \langle 1 - e \rangle$ and $C_2' = \langle 1 - \mu_{-1}(e) \rangle$.

It is just as easy to demonstrate the converse.   Q.E.D.

THEOREM 5 [10]. $\overline{C}$ of length $n + 1$ is an extended cyclic, self-dual code over $GF(q)$ $(q = p^i)$ iff C is an odd-like duadic code with splitting given by μ_{-1} and $n \equiv -1 \pmod{p}$.

This theorem tells us for which n extended cyclic, self-dual codes of length $n + 1$ can exist over $GF(q)$. First, n has to be such that duadic codes exist, i.e., q must be a square (mod n). Secondly, μ_{-1} must give a splitting for n. This always occurs when the order of q (mod n) is odd. If $q = 2$, all extended cyclic, duadic codes are self-dual if $n = \prod p_i^{a_i}$ where each p_i is a prime $\equiv -1 \pmod{8}$. Some are self dual if $n = \prod p_i^{a_i}$ where the order of 2 (mod p_i) is odd when $p_i \equiv 1 \pmod{8}$.

Note that we know that every even-like binary duadic code in our example of length 31 is self-orthogonal. Also every extended odd-like duadic code of this length is self-dual. Further we can easily compute the idempotents and hence a generating matrix for every cyclic binary self-orthogonal (31, 15) code and also for every extended cyclic binary self-dual (32, 16) code.

For the ternary example, we know that there are self-orthogonal (13, 6) codes and in fact they are precisely the even-like duadic (13, 6) codes. Constructing their idempotents is far more complicated.

We can even tell something about weights in self-orthogonal binary, cyclic codes. For length 31 we see that the generating idempotents of the self-orthogonal (31, 15) cyclic codes have weight 16. It is not difficult to show from this that the weights of all vectors in these codes must be divisible by 4, i.e. these codes are doubly-even.

THEOREM 6. *If C is a self-orthogonal, binary $\left(p, \frac{p-1}{2}\right)$ cyclic code for p a prime, then the weights of all vectors in C are divisible by 4. The vectors in C are all the even weight vectors in $C^{\perp}$. All odd weights of vectors in $C^{\perp}$ are $\equiv p \pmod{4}$.*

If $\overline{C}$ is an extended cyclic binary self-dual code of length $p+1$ for p a prime, then p must be $\equiv -1 \pmod 8$. Also all weights of vectors in $\overline{C}$ are divisible by 4.

PROOF This is easily determined by computing the weight of the idempotent. For example when $p \equiv -1 \pmod 8$ the even weight idempotent has weight $\frac{p-1}{2} + 1 = \frac{p+1}{2}$ which is divisible by 4. A generating matrix of the even-weight code is given by this idempotent and its cyclic shifts. As the code is self-orthogonal any linear combination of these vectors must have weight divisible by 4.   Q.E.D.

It is often useful to know when cyclic self-orthogonal codes of any dimension cannot exist.

THEOREM 7. *If $p^i \equiv -1 \pmod n$ for some integer i, then there are no cyclic self-orthogonal codes of length n over $GF(p)$.*

PROOF Suppose $C = \langle e \rangle$ is self-orthogonal. Then $C \subset C^{\perp} = \langle 1 - \mu_{-1}(e) \rangle$. Now $e\left(1 - \mu_{-1}(e)\right) = e$ since $1 - \mu_{-1}(e)$ is the multiplicative unit of $C^{\perp}$. Hence $e\mu_{-1}(e) = 0$. Because $-1 \equiv p^i \pmod n$ and $\mu_p(e) = e$, we have $\mu_{-1}(e) = e$ so that $e^2 = e = 0$.   Q.E.D.

See [11] for further results along this line of reasoning.

If n is a prime, then the condition that $p^i \equiv -1 \pmod n$ is the same as the condition that p has even order $\pmod n$. This latter means that there is no self-orthogonal $\left(n, \frac{n-1}{2}\right)$ cyclic code over $GF(p)$. In fact it can be shown in general that if there does not exist a cyclic, self-orthogonal $\left(n, \frac{n-1}{2}\right)$ code then there does not exist a cyclic self-orthogonal code of length n and any dimension. This is the meaning of the next theorem.

THEOREM 8. *If there is a divisor w of n and an integer i so that $p^i \equiv -1 \pmod w$, then there are no cyclic self-orthogonal codes of length n over $GF(p)$.*

5. Implications for Cyclic Difference Sets.

A *symmetric* (v, k, λ) *design* is a set v points and a set D of subsets of size k of these points called *blocks* with the following properties. We suppose $k > \lambda$.

a) Every 2 points are contained in exactly λ blocks and

b) Every 2 blocks meet in eactly λ points.

These properties imply [5] that there are v blocks and that every point is on the same number of blocks. It is common to set $n = k - \lambda$ and call n the *order* of the design.

A central problem in symmetric designs or combinatorial designs in general is the question of the existence of a design with specified parameters. This is often a quite difficult problem. A problem which is often more manageable is the existence of a design with specified parameters and a certain type of group acting on it. Usually one starts with a cyclic group.

The *group* of a *design* is the set of all permutations of the points which send blocks onto blocks. A symmetric design with a regular automorphism group and a difference set are essentially the same.

When $\lambda = 1$, a symmetric design is called a *projective plane* and there have been great efforts to determine for which orders a plane exists.

A relatively new tool in settling these questions is the code generated by the design. In order to obtain this we write the *incidence matrix* of the design. If D is a symmetric (v, k, λ) design, this is the $v \times v$ matrix whose rows are the blocks of the design, that is we represent each block by a length v binary vector with a 0 for a point not on the block and a 1 for a point on the block. We can then consider the code generated by the rows of this matrix over some field. We choose the field so that we can say something about the design generated code.

If we consider the $(7, 3, 1)$ symmetric design which is also the projective plane of smallest order, 2, then the following is an incidence matrix.

0	1	2	3	4	5	6
0	1	1	0	1	0	0
0	0	1	1	0	1	0
0	0	0	1	1	0	1
1	0	0	0	1	1	0
0	1	0	0	0	1	1
1	0	1	0	0	0	1
1	1	0	1	0	0	0

The set $\{1, 2, 4\}$ is a $(7, 3, 1)$ difference set in the (additive) group of integers (mod 7) since

$$1 - 2 \equiv 6 \qquad 2 - 1 \equiv 1 \qquad 4 - 1 \equiv 3$$
$$1 - 4 \equiv 4 \qquad 2 - 4 \equiv 5 \qquad 4 - 2 \equiv 2$$

We can also recognize these vectors as the vectors of weight 3 in the Hamming $(7, 4, 3)$ code thus realizing that the dimension of the binary code generated by this design is 4, its minimum weight is 3 and the vectors of weight 3 in the code are precisely the blocks of the design or lines of the plane. This is a nice situation indeed.

If we have a projective plane of even order n, then the rows of its $(n^2 + n + 1) \times (n^2 + n + 1)$ incidence matrix have odd weight $n + 1$ and every two rows have exactly one 1 in common. Hence the extended code of length $n^2 + n + 2$ is self-orthogonal so its dimension and the dimension of the original code are both less than or equal to $\frac{n^2 + n + 2}{2}$. When $n \equiv 2 \pmod 4$, it can be shown to be exactly equal to $\frac{n^2 + n + 2}{2}$ [5]. Indeed we can use this fact and the theorems of the last section to give a simple coding theory proof of the non-existence of certain cyclic projective planes.

THEOREM 9 [9]. *The only cyclic projective plane P of order $n \equiv 2 \pmod 4$ is the projective plane of order 2.*

PROOF Let C be the binary cyclic code of length $n^2 + n + 1$ generated by the lines of P. As remarked above $\overline{C}$ is self-orthogonal so that $\dim C \leq \frac{n^2+n+2}{2}$. Since $n \equiv 2 \pmod 4$, $\dim C = \frac{n^2+n+2}{2}$. It is fairly easy to show that $C = \langle h \rangle + C^{\perp}$ and that $C^{\perp}$ is a self-orthogonal $(n^2 + n + 1, \frac{n^2+n}{2})$ code. By Theorem 4 then $C^{\perp}$ and hence C are duadic codes.

From our previous discussion we know that the weight of C's generating idempotent is $\frac{n^2+n}{2}$. It is also known [10] that the weight of the lines, $n + 1$, is the minimum weight of C and that these lines are the only vectors of that weight in C. As μ_2 sends C onto itself it must send the vectors of weight $n + 1$ onto themselves. Since μ_2 has a fixed point, 0, it must have a fixed line, x. Hence $\mu_2(x) = x$ which shows that x is an idempotent and is the generating idempotent of C by its construction. Hence its weight, $n + 1$, must equal $\frac{n^2+n}{2}$. The only positive solution of this equation is $n = 2$.   Q.E.D.

Clearly by Theorem 9 there is no cyclic projective plane of order 10. The non-existence of this plane can also be shown easily with Theorem 8. A cyclic projective plane of order 10 would be a cyclic $(111, 11, 1)$ symmetric design. Let C be the code generated by the set of $\{\ell_i + \ell_j$ where ℓ_i and ℓ_j are lines$\}$. Each $\ell_i + \ell_j$ has weight 20 and it is simple to show that any $\ell_i + \ell_j$ is orthogonal to any other sum of lines. Hence C is a self-orthogonal code. Theorem 8 rules out such a C as $2 \equiv -1 \pmod 3$. Invocation of Theorem 8 rules out many parameter sets (v, k, λ) even when λ is greater than 1 [5].

This type of approach can be generalized to abelian symmetric designs or difference sets with respect to an abelian group [15].

Other Applications. The connections between coding theory and number theory are too numerous to mention. So only the ones most familiar to me are noted here, this has no implication about the importance of the topics either mentioned or omitted.

It is important to know when codes are considered the same. We say that two codes are *equivalent* (or permutation equivalent), if there is a coordinate permutation which sends one code onto the other. It is known [4] that if two cyclic codes of prime length are equivalent, then they must also be equivalent by a multiplier. This is very convenient, especially for binary codes, as we can determine the idempotents easily. As a multiplier sends a cyclic code onto another cyclic code iff it sends the idempotent of one to the idempotent of the other, we can determine equivalent cyclic codes by this process. In [4] we use a result of Palfy [8] to show that two cyclic codes of length n are equivalent iff they are multiplier equivalent whenever $\gcd(n, \phi(n)) = 1$ or $n = 4$ or $n = pr$, $p > r$ are primes and the Sylow p-subgroup of the automorphism group of C has order p.

It is possible to consider *generalized multipliers* of cyclic codes of length p^m. When $m = 2$, this is the mapping

$$\mu_d : k \to id \pmod p + pj \quad \text{where} \quad k = i + pj \quad \text{for} \ \ 0 \leq i, \ j < p.$$

As usual p is a prime. Let q be a prime power relatively prime to p and let z be the greatest integer such that $p^z | (q^t - 1)$ where $t = \text{ord}_p(q)$. In the case $z = 1$, generalized multipliers send cyclic codes onto cyclic codes and also generating idempotents onto generating idempotents [4]. It is interesting that cyclic codes of length p^2 are equivalent iff they are equivalent by either a multiplier or a multiplier times a generalized multiplier [4].

A *divisible code* is a code where all the weights of codewords share a common divisor larger than one [17]. We have seen this in binary doubly-even codes. Repetition codes are essentially the only divisible codes for divisors prime to the field size. Ward [17] devises a divisibility test on a spanning set when the divisor is a power of the characteristic. To carry out his computations he lifts the words of a spanning set of a code over a finite field up to words in a p-adic field. There the weight enumerator can be given in terms of functions giving Teichmüller representatives. Polarizing formulas for these allow the divisibilities to be checked.

The many relations between codes and lattices should also be mentioned. There are many ways to obtain a code over F_p from a lattice L. If L is contained in Z^m, the simplest way is the reduction (mod p) of L. Lander [5] generalizes this as follows. He lets $\pi^m : z^m \to \overline{F}_p^m$ be the "reduction (mod p)" homomorphism and he defines the F_p-codes:

$$C_\alpha = \pi^m(p^{-\alpha}L \cap Z^m)$$

for $\alpha = \dots, -2, -1, 0, 1, 2, \dots$

This gives a set of nested codes

$$\dots \subseteq C_{-1} \subseteq C_0 \subseteq C_1 \subseteq C_2 \subseteq \dots$$

Lander is able to determine the dimensions of these codes and identify duals with respect to a non-degenerate scalar product defined on Q^m. Under certain conditions one of these codes is self-dual.

John Conway and Neil Sloane have written a marvelous book, "Sphere Packings, Lattices and Groups" [2] which contains a wealth of information about the connections of these objects to various classes of codes and many specific codes. Many of the densest lattice packings are obtained from error correcting codes. Several construction methods are given to obtain lattices from codes. They find the strongest connections between self-dual codes and self-dual lattice packings and the strongest connections of all when they compare doubly-even self-dual codes with self-dual lattices in which the norm of every vector is even. In the latter case they give parallel theorems concerning upper and lower bounds on the best codes and lattices and parallel theorems characterizing the weight enumerator of the code and the theta series of the lattice.

B.C.H. codes are a popular class of cyclic codes which are used in many practical situations. Given a distance d, length n and finite field $GF(q)$ one can construct a B.C.H. code of length n over $GF(q)$ with minimum weight d or more. Only weak bounds on the dimension of the constructed code are known in general. This construction was generalized to Goppa codes. The further generalization of this construction by Goppa [3] and Tsfasman, Vladut and Zink [16] created quite a stir in the coding community as they demonstrated the existence of better codes than had been known before albeit over large fields, $GF(49)$ or larger. They constructed codes from algebraic curves over finite fields. Their work led to the question of the maximum number of points on an algebraic curve of genus g. The Riemann-Roch theorem is a major tool in determining the dimension of the code constructed. This is a very large topic and we just refer the reader to [6, 7].

REFERENCES

1. R.A. Brualdi and V. Pless, Polyadic codes, *Discrete Applied Math.* **25** (1989), 3-17.

2. J.H. Conway and N.J.A. Sloane, Sphere Packings, Lattices and Groups (Springer-Verlag, New York, 1988).

3. V.D. Goppa, Algebraico-geometric codes, *Math. USSR-Izv.* **21** (1983), 75-91.

4. W.C. Huffman, V. Job, and V. Pless, Multipliers and generalized multipliers of cyclic objects and cyclic codes, to appear in *J.Comb. Th.(A)*.

5. E.S. Lander, Symmetric Designs: An Algebraic Approach (Cambridge University Press, New York, 1983).

6. J.H. van Lint and G. van der Geer, Introduction to Coding Theory and Algebraic Geometry (Birkhäuser Verlag, Boston, 1988).

7. C. Moreno, Editor, algebraic Curves over Finite Fields (Cambridge University Press, New York, 1990).

8. P.P. Pálfy, "Isomorphism problem for relational structures with a cyclic automorphism," *Eur. J. Comb. Th.* **8** (1987), 35-43.

9. V. Pless, "Cyclic projective planes and binary, extended cyclic self-dual codes," *J. Comb. Th.(A)* (1986), 331-333.

10. V. Pless, "Duadic codes revisited," *Congressor Numerantium* **59** (1987), 225-233.

11. V. Pless, "Cyclic codes and cyclic configurations," *Contemporary Mathematics* **111** (1990), 171-177.

12. V. Pless, "Introduction to the Theory of Error-Correcting Codes" (2nd edition, John Wiley and Sons, New York, 1989).

13. V. Pless, J.M. Masley, and J.S. Leon, "On weights in duadic codes," *J. Comb. Th.(A)* (1987), 6-21.

14. V. Pless, and J.J. Rushanan, "Triadic Codes," *Lin. Algebra and Its Appl.* **98** (1988), 415-433.

15. J.J. Rushanan, "Duadic codes and difference sets," to appear in *J. Comb. Th.(A)*.

16. M.A. Tsfasman, S.G. Vladut and I. Zink, "Modular curves, Shimura curves and Goppa curves, better than Varshamov-Gilbert bound," *Mach. Nochr.* **109** (1982), 21-28.

17. H.N. Ward, "Weight polarization and divisibility," *Discrete Math.* **83** 315-326, 1990.

Proceedings of Symposia in Applied Mathematics
Volume 46, 1992

Number Theory in Computer Graphics

M. DOUGLAS McILROY

ABSTRACT. Computer graphics is geometry on a grid. Hence elementary number theory plays a central role in the design and analysis of basic curve-drawing algorithms. Circles involve matters of representability by sums of squares; straight lines involve continued fractions.

1. Introduction

Computers make drawings by coloring pixels in a bitmap, which may be thought of as the points of a plane integer lattice. People made digital pictures for thousands of years before computers and well before number theory, too (Figure 1). But with computers, where algorithm supplants artistry, the mathematics becomes more important. Drawing a figure becomes a problem in two-dimensional Diophantine approximation: picking points of the lattice to get a good fit.

Figure 1. A bitmap image from the Chicama Valley, Peru, circa 2000 BC. American Museum of Natural History. Reprinted with permission from G. G. S. Bushnell, *Ancient Arts of the Americas*, Thames and Hudson, London, 1967.

1991 *Mathematical subject categories:* Primary 11-01; Secondary 68U05, 11J70, 11B57, 11J06.
This paper is in final form. No version of it will be submitted for publication elsewhere.

© 1992 American Mathematical Society
0160-7634/92 $1.00 + $.25 per page

Resolutions in digital images are typically around one part in a thousand— even coarser on small personal-computer screens. Roundoff becomes visible, and often annoying. Single-pixel errors can have disastrous effects (Figure 2). The details matter, and that is where number theory comes in. Continued fractions, for example, explain the details of the "jaggies" that appear in diagonal lines. Farey series, too, enter into the characterization of digitized lines. The existence and nonexistence of solutions to quadratic Diophantine problems bears on the uniqueness of digital images of circles and ellipses.

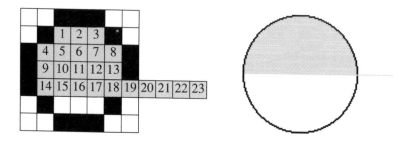

Figure 2. Costly roundoff. (Left) A digital circle of radius 3 with a single-pixel gap, as might result from roundoff error. An attempt to fill it by "painting" the interior pixels in the numbered order fails; the grey paint leaks out. (Right) A similar gap in a digital circle of radius 50 is almost invisible, but leaks the same way.

2. Freeman approximation

In drawing a digital curve, we wish to select pixels that are close to the continuous curve it approximates. Among various criteria for judging closeness [11], one due to Freeman has particularly nice properties [14]. It is invariant under all symmetry operations of the integer lattice: integer translations, half turns, quarter turns, and reflections. And it reduces the two-dimensional problem to a collection of one-dimensional problems.

Definition 0. Unless otherwise specified, *point* means a point of the plane integer lattice.

Definition 1. The *Freeman approximation* to a curve is the set of points (x,y) for which the curve intersects either of the closed unit segments, $H(x,y)$ and $V(x,y)$ centered on (x,y), where

$$H(x,y) = \{(u,y) \mid x - \tfrac{1}{2} \leq u \leq x + \tfrac{1}{2}\},$$

$$V(x,y) = \{(x,v) \mid y - \tfrac{1}{2} \leq v \leq y + \tfrac{1}{2}\}.$$

Freeman approximation picks, for each intersection of the curve with a grid line of the lattice, the point nearest to the intersection. Figure 3 shows Freeman points as dots and the segments $H(x,y)$ and $V(x,y)$ as bars.

Ideally a digital curve should look like a curve. It should appear uniformly thin and connected. It should not be too jagged.

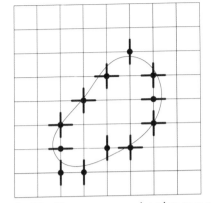

Figure 3. The Freeman approximation to a curve.

Definition 2. A set S of points is *connected* if and only if every pair of points in S can be joined by a path of king moves* within S.

The Freeman approximation to a connected curve is necessarily connected.

Definition 3. A differentiable curve is said to be *predominantly horizontal* if its slope is in the range $[-1,1]$ and *predominantly vertical* if its slope is in the range $[-\infty,-1] \cup [1,\infty]$.

Definition 4. A set of lattice points is *thin* if and only if at most two points of the set are incident on any unit square in the lattice.

The top part of the curve in Figure 4 is predominantly horizontal. It is easy to see that the Freeman approximation to a predominantly horizontal curve must be thin except where it passes exactly halfway between vertically adjacent lattice points.

Thinness is a sometime thing, which can be defeated in several ways, as shown in Figure 4. Some failures of thinness may be defined away. When a curve passes exactly half way between two adjacent lattice points, we may be able to break the tie by an arbitrary choice. When a curve doubles back on itself, we can save the appearances by thinking of the several branches of the curve as lying in different sheets. However, when a curve switches between predominantly horizontal and predominantly vertical there may occur a *square corner*, where the approximation includes three corners of a grid square. This last kind of departure from thinness seems inescapable.

Thinness and connectedness are preserved under the symmetry operations on the lattice.

Half-Freeman approximations are common in computer graphics. A half-Freeman approximation picks points on grid lines that run in only one direction. If the curve is predominantly horizontal, points are picked on vertical grid lines and vice versa (Figure 5). It is easy to see from the figure that the Freeman points on

---

* In chess, a king move changes one or both coordinates of a point by ± 1.

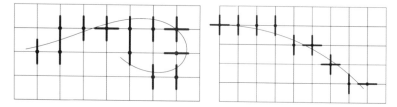

Figure 4 (Left). Thinness is defeated (1) if a curve passes exactly half way between two points as near the left end, (2) if the curve doubles back on itself as near the middle, or in some instances (3) if the curve changes between predominantly horizontal and predominantly vertical, making square corners as at the right side of the figure.

Figure 5 (Right). Vertical bars mark the vertical half-Freeman approximation to a predominantly horizontal octant of a circle. Horizontal bars marking the horizontal half-Freeman approximation agree with the vertical bars except possibly at an end point.

horizontal grid lines coincide with the half-Freeman points on vertical grid lines—except possibly for single outlying points at the ends of the curve.

Half-Freeman approximations to a connected curve are necessarily connected. They are also thin, or can be made so by breaking ties. (To preserve connectedness, ties may have to be broken in a consistent direction.) Thus half-Freeman approximations make visually convincing curves. They are usually easy to compute, too, so they have become ubiquitous in computer graphics. Circles, for example, are usually drawn as eight half-Freeman octants (see box).

To preserve connectivity under half-Freeman approximation, a curve must be split into predominantly horizontal and predominantly vertical sections. The section joins may need special treatment. When a circular quadrant is approximated by octants, there may be one point that is not a half-Freeman point of either octant (Figures 5 and 6).

3. Circles

Circles enter number theory through the study of the Diophantine equation

$$x^2 + y^2 = r^2. \tag{1}$$

In computer graphics, where coordinates are naturally integral, it is customary—and computationally convenient—to limit attention to just such a *standard circle* centered at (0,0) with radius r, or possibly r^2, taken to be integral. The Freeman approximation to the first quadrant is the set of lattice points (x,y) that satisfy either of the inequalities

$$\left| y - \sqrt{r^2 - x^2} \right| \le 1/2, \quad 0 \le x \le r \tag{2a}$$

$$\left| x - \sqrt{r^2 - y^2} \right| \le 1/2, \quad 0 \le y \le r \tag{2b}$$

How to draw a circle

Formula (3) is the basis of a neat program to trace one octant of a circle [12], [17]. The other octants can be filled in by symmetry.

Start at $(x_0, y_0) = (0, r)$ and then step along the circle computing further points (x_i, y_i) by the scheme

$$x_{i+1} = x_i + 1$$

$$y_{i+1} = \begin{cases} y_i - 1, & \text{if } (y_i - 1/2)^2 + x_{i+1}^2 - r^2 > 0 \\ y_i, & \text{otherwise.} \end{cases}$$

Stop when x_{i+1} would exceed y_{i+1}. The last point drawn may fall beyond the end of the octant. In this event, the point can be shown to be a Freeman point anyway, as in Figure 6.

The scheme avoids any computation of square roots. The single fraction can be eliminated, leaving only integer calculations. Finite differences may be used to update the the quadratic test at each step, thus reducing the calculation of a circular octant to just integer additions, subtractions, and comparisons. The final algorithm is so easy that it comes built into many graphics devices.

Inequality (2a) is satisfied by the greatest integer y such that

$$(y - 1/2)^2 \leq r^2 - x^2 \tag{3}$$

The half-Freeman approximation to an octant of a standard circle is necessarily thin because a solution of (3) with $y > 1/2$ is unique. If it were not, then the circle would pass exactly half way between some pair of adjacent lattice points, say (x, y) and $(x, y + 1)$, making

$$y + 1/2 = \sqrt{r^2 - x^2} \ .$$

The left side of this equality is rational, but not integral. The right side is the square root of an integer, and therefore either irrational or integral. Hence the situation is impossible and the approximation is thin.

3.1. Other approximation criteria. Freeman circles are remarkably robust, in the sense that different approximating criteria lead to the same approximations. Three natural measures of closeness are

Distance. Euclidean distance from the lattice point to the curve.

Displacement. Distance from the lattice point to the curve measured along the grid line. (This is the Freeman approximation.)

Residual. The absolute value of a defining function $f(x, y)$; the curve is the solution set of $f(x, y) = 0$.

For a standard circle, the minimum-distance approximation picks on each grid line a point (x, y) to minimize $|\sqrt{x^2 + y^2} - r|$. The minimum-displacement approximation minimizes $|y - \sqrt{r^2 - x^2}|$ on vertical grid lines, and a transposed

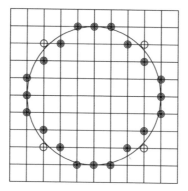

Figure 6. A Freeman circle of radius 4. The black dots mark half-Freeman points of their respective octants. The open dots are Freeman points that are not half-Freeman points of any octant; these points are square corners.

formula on horizontal grid lines. The minimum-residual approximation minimizes $|x^2 + y^2 - r^2|$. Surprisingly, all three approximations are the same [4], [24], [25].

Theorem 1. If r^2 is integral, the minimum-distance, minimum-displacement, and minimum-residual approximations to the circle $x^2 + y^2 = r^2$ are unique and coincide. The approximations are thin except possibly for square corners on the lines $y = \pm x$.

The proof is elementary, but involves considerable case analysis. One case will give some of the flavor. Suppose (x,y) is a minimum-residual point outside the circle and $(x, y-1)$ is a minimum-distance point inside. Both x and y are positive. In analytic terms,

$$r^2 - (x^2 + (y-1)^2) \geq (x^2 + y^2) - r^2,$$
$$r - \sqrt{(x^2 + (y-1)^2)} \leq \sqrt{x^2 + y^2} - r.$$

Both sides of both inequalities are positive. Hence the inequalities may be divided to obtain

$$r + \sqrt{x^2 + (y-1)^2} \geq \sqrt{x^2 + y^2} + r,$$

which is absurd. The supposed configuration is impossible.

We originally set out to approximate circles of integer radius. Theorem 1 covers somewhat more: circles with radius equal to the square root of an integer. In particular the three approximations must agree for any circle centered on a lattice point and passing through some lattice point.

It is natural to consider admitting integer diameters and half-integer centers. One could then, for example, cleanly approximate circles inscribed in squares of arbitrary integer dimensions. In this widened class the guarantee of thinness disappears and the approximations may disagree pairwise. Taken together, though, the three approximations always vote unanimously for a unique answer [25].

Theorem 2. If $2x_0$, $2y_0$, and $(2r)^2$ are integral, the intersection of the minimum-distance, minimum-displacement, and minimum-residual approximations to a quadrant of the circle $(x - x_0)^2 + (y - y_0)^2 = r^2$ is nonempty and unique on any grid line that intersects the quadrant.

Theorem 2 is best possible, in the sense that half integers cannot be replaced by any finer rational subdivision of the lattice. If $q > 2$, there exist circles with integral qx_0, qy_0, and $(qr)^2$ for which there is not even majority consensus. The intersection of any two of the three approximations to these circles is disconnected [25].

3.2. Square corners. Just how often in approximating a standard circle of integral radius will we meet square corners as in Figure 6? The answer—about twice between successive powers of 34—depends on the Pell equation [23], [25].

Theorem 3. Square corners appear in the Freeman approximation to a standard circle of integral radius r if and only if r = 4, 11, 134, 373, 4552, 12671, 154634, 430441, ⋯ .

The sequence of radii in Theorem 3 satisfies the linear recurrence $r_{k+2} = 34r_k - r_{k-2}$.

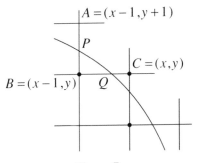

Figure 7.

Figure 7 shows the conditions under which a square corner can happen. The square corner C must be on the diagonal, where $x = y$, and we must have $PB < PA$ and $QC < QB$. By Theorem 1, we may work in terms of residuals, and avoid square roots. Point B is inside the circle; the residual there, $(x - 1)^2 + y^2 - r^2$, is negative. The inequality $PB < PA$ implies that the residuals at B and A satisfy

$$-((x - 1)^2 + y^2 - r^2) < (x - 1)^2 + (y + 1)^2 - r^2,$$

and similarly at C and B,

$$x^2 + y^2 - r^2 < -((x - 1)^2 + y^2 - r^2).$$

Set $x = y$ in these inequalities and simplify, to find

$$4x^2 - 2x + 1 < 2r^2 < 4x^2 - 2x + 3,$$

which, because r is integral, is equivalent to

$$r^2 = 2x^2 - x + 1. \tag{4}$$

Completing the square in (4) and eliminating fractions gives

$$8r^2 = (4x - 1)^2 + 7. \tag{5}$$

The solutions of (5) are solutions of the Pell equation $2p^2 - q^2 = 7$ with p even and $q \equiv 3 \pmod 4$. Standard methods of solution [29] over the unique factorization domain $\mathbf{Z}(\sqrt{2})$ lead to Theorem 3 and the associated recurrence.

One way to read (4) is that the residual at C, namely $x^2 + x^2 - r^2$, is $x - 1$. It follows that the residuals at B and A are $-x$ and $x + 1$. The magnitudes of the three residuals are contiguous integers; the conditions for a square corner are delicate indeed.

3.3. Ellipses, etc.

Algorithms for approximating standard ellipses resemble those for circles [12], [27]. Again, as with standard circles, the Freeman approximation to an ellipse with semiaxes of integral length is unique. In other words, it is impossible for the ellipse

$$\frac{x^2}{a^2} + \frac{y^2}{b^2} = 1$$

with integral a and b to pass exactly half way between adjacent lattice points. To prove it, suppose that the ellipse passes through a point, $(x, z/2)$, where z is an odd integer. We may assume that $\gcd(x, a) = \gcd(z, b) = 1$; if not, we could reduce the fractions in the defining equation

$$\frac{x^2}{a^2} + \frac{z^2}{4b^2} = 1$$

to get a counterexample of the same form where the assumption does hold. The sum of two fractions in lowest terms can be 1 only if their denominators are the same. Hence $a = 2b$. Because x/a is in lowest terms, x must be odd. Consequently (x, z, a) is a Pythagorean triple with parities (odd,odd,even) that satisfies

$$x^2 + z^2 = a^2.$$

As is well known, no such triple exists, for that would imply

$$1 + 1 \equiv x^2 + z^2 \equiv a^2 \equiv 0 \pmod 4.$$

Standard ellipses with integer semiaxes are about the most complicated curves for which a fully satisfactory understanding is available in computer graphics. Other conics, and even nonstandard circles, are customarily handled in an ad hoc way. There's plenty of room to exploit other knowledge from number theory. Here are some questions.

1. If arbitrary centers, (x_0, y_0), and radii are allowed, how many distinct digital circles of radius less than r exist? To what extent does the answer depend on the approximating criterion? Characterize the equivalence classes that digital approximation induces in (x_0, y_0, r) space.

2. Recognize digital circles efficiently.

3. How rough is a digital curve?

4. Describe properties of other digital curves. General 2nd-degree and 3rd-degree polynomials (conics and splines) are of particular interest in computer graphics.

5. The union of digital circles of all integer radii does not cover the integer lattice. Describe the gaps.

4. Straight lines and chain codes

The path of a digital curve is often recorded as a *chain code,* or differential encoding of pixel-to-pixel moves [31]. The intriguing structure of chain codes of straight lines eluded satisfactory explanation until a connection was made with continued fractions.

The chain-code representation of a general digital curve is relatively compact, certainly much more so than a sequence of coordinate pairs. Fortuitously in two dimensions there are 8 directions to nearest neighbors, so general chain codes fit perfectly in 3 bits. Here we shall adopt a limited definition, with just two instead of 8 directions of motion, adequate for the analysis of digital straight lines of positive or zero slope [6].

Definition 5. A *chain code* of a monotone nondecreasing curve, $f(x)$, is a string of 0's and 1's. Let i be an integer abscissa and let $h(i) = \lfloor f(i+1) \rfloor - \lfloor f(i) \rfloor$, where $\lfloor \ \rfloor$ is the integer part function. Then the chain code can be parsed into words, $01^{h(i)}$, made of a single zero followed by $h(i)$ ones.

In the chain code of a predominantly horizontal curve, 1's cannot occur contiguously.

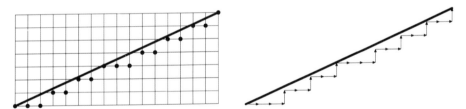

Figure 8. (Left) A digital segment of slope 7/16. (Right) The chain code of the same segment.

A chain code designates a Manhattan path (Figure 8). The digital line segment in Figure 8 has slope $m = 7/16$. If the lower left corner is taken to be $(0,0)$, then the ordinate y of each point (x,y) of the digital segment is the largest integer y that satisfies $y \leq mx$. In other words, the points form a greatest lower bound integer approximation to the line,* sometimes called the *spectrum* of m [15]. The chain code is an infinite repetition of the string

$$0001001001000100100100 1.$$

---

* The greatest lower bound approximation is the same as the Freeman approximation of the shifted line $y = mx - 1/2$, with ties broken in favor of the upper point.

A predominantly horizontal line has a chain code with at least one 0 for every 1. The 1's are isolated and spread as uniformly as possible among the 0's. Thus between every two successive 1's there must be p or possibly $p+1$ zeros, for some integer p. Continued fractions provide the details of the distribution of 1's.

Chain codes by other names have a long history in the geometry of numbers. The "cutting sequence" of grid lines by a line that passes through no lattice points is precisely the chain code, where 0 denotes a vertical grid crossing and 1 a horizontal one. Cutting sequences, presented as the sequence of chain-code exponents, p or $p+1$, also arise in counting the integers between successive terms of the series $\{i\alpha\}$ of integer multiples of an irrational α. In these guises, the study of chain codes goes back at least a century to Christoffel and Markoff [32].

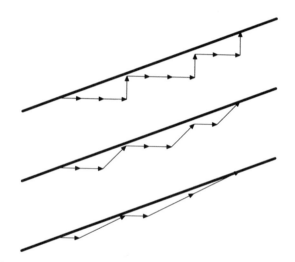

Figure 9. The chain code for a line of slope 3/8 in the standard basis (top) is 00010001001; in the basis transformed by S it is 00100101 (middle), and transformed by S^2 (bottom) it is 01011.

4.1. Chain code transformations and continued fractions.
Bruckstein defined a set of invertible transformations among chain codes of digital lines [6]. Two of them, called X and S, are given here.

X. Parse the string into words 0 and 1. Exchange 0 and 1.

S. Parse the string into words 0 and 01. Replace 01 by 1.

Theorem 4. The result of applying either of the transforms X or S to a string of 0's and 1's is the chain code of a digital line if and only if the original string is the chain code of a digital line.

Each transform can be explained as a change of basis in the lattice, and is thus associated with a matrix. Transform X transposes the lattice; it has matrix M_X:

$$M_X = \begin{bmatrix} 0 & 1 \\ 1 & 0 \end{bmatrix}.$$

Transform S takes as a new basis the vectors originally coded as 0 and 01 (Figure 9). Code 0, a (1,0) step in original coordinates, maps into code 0 or a (1,0) step in the new. Code 01, a (1,1) step in original coordinates, maps into code 1 or a (0,1) step in the new. The matrix associated with S is thus

$$M_S = \begin{bmatrix} 1 & 1 \\ 0 & 1 \end{bmatrix}^{-1} = \begin{bmatrix} 1 & -1 \\ 0 & 1 \end{bmatrix}.$$

When transform S is iterated p times, the short and long words, 0^p1 and $0^{p+1}1$, reduce to 1 and 01 respectively. The resulting string is not parsable into 0's and 01's, so the transform is no longer applicable. The iterated transform has the matrix

$$M_{S^p} = (M_S)^p = \begin{bmatrix} 1 & -1 \\ 0 & 1 \end{bmatrix}^p = \begin{bmatrix} 1 & -p \\ 0 & 1 \end{bmatrix}.$$

Continued fractions arise when the transform XS^p is applied repeatedly [5]. (Since S^p changes the predominant symbol from 0 to 1, transform X must be used after each S^p to restore the predominancy of 0.) Table 1 shows the process applied to the chain code for Figure 8. In the table, the operation "shift" simply changes viewpoint on the periodic chain code of an infinite line. The values of p in the table are the partial quotients, [2,3,2], of the continued fraction for the slope [16]:

$$\frac{7}{16} = \cfrac{1}{2 + \cfrac{1}{3 + \cfrac{1}{2}}}.$$

Intuitively, the approximation must rise 7 steps in a run of 16. ("Rise" and "run" are used here in the engineering sense, meaning horizontal and vertical changes.) A rise of 1 in a run of 2 almost does it, so 2 is the first partial quotient. To this degree of approximation, the chain code is 001. But the approximation is too steep, rising 7 steps in a run of 14, not 16 as desired. Thus two out of every seven 001 words must be lengthened. We need one extra 0 in every 3½ words. And that we do by lengthening one word in every 3 (the next partial quotient) to make a superword 0001001001, and sticking in one 001 word after each 2 (the last partial quotient) superwords.

Algebra explains better than English. Since the chain code of a predominantly horizontal line of slope m has one 1 for every p or possibly $p+1$ zeros, we have $p = \lfloor 1/m \rfloor$. Then

$$m = \frac{1}{1/m} = \frac{1}{\lfloor 1/m \rfloor + (\text{number less than 1})},$$

Thus we see that the iteration exponent p is in fact the partial quotient in the continued fraction for m. The continued fraction for the residual "number less than 1" corresponds to the chain code as transformed by XS^p.

In every case, the transforms must convert invertibly from integer coordinates to integer coordinates. Hence all the matrices must have integer entries and

Table 1. Transforming the chain code for slope 7/16.

Sequence	Applicable transform
000100100100001001001001	$S^p, p = 2$
011101111	X
100010000	Shift
000010001	$S^p, p = 3$
011	X
100	Shift
001	$S^p, p = 2$
1	X
0	

determinant ± 1. Two-by-two matrices of this type form a "general linear group," customarily denoted $GL(2,\mathbf{Z})$. The group elements represent all possible basis changes in the lattice [10]. Transforms X, S, and one other, in particular

$$\begin{bmatrix} -1 & 0 \\ 0 & 1 \end{bmatrix}, \tag{6}$$

generate the group. This group gives a full account of invertible chain-code transformations. As the whole group includes basis changes that turn lines of positive slope into lines of negative slope, it requires a general domain of chain codes, where negative as well as positive steps are admitted. Transform (6) maps chain codes in our limited subdomain into chain codes outside the subdomain, as do higher powers of S than S^p.

4.2. Association of segments and chain codes. By Theorem 4, the method of Table 1 can be used to test whether a periodic or finite sequence of 0's and 1's is the chain code of a digital line or segment. A string of 0's and 1's is a chain code (or period thereof) for a line of rational slope if and only if the the string reduces to all 0's in a finite number of S^p and X steps [33]. The number of such steps is $O(\log n)$, the same as for the continued fraction algorithm [20], so the overall time to decide by this method whether a sequence of n 0's and 1's form the chain code of a straight line is $O(n \log n)$. (This running time is not optimal; the straightness of a chain code can be decided in time $O(n)$ [3].)

The chain code of a given segment is unique. Moreover the chain codes of all segments of a given rational slope with a common projection on the x axis are unique up to a cyclic shift. But many segments may have the same chain code. For any finite chain code C containing n zeros, there is an equivalence class of pairs (m,b) for which the segment

$$y = mx + b, \quad 0 \le x \le n \tag{7}$$

has chain code C.

Let us find the shape of the equivalence sets. If the segment (7) passes between, but does not touch, the lattice points (x,y) and $(x,y+1)$, then b lies in the open interval $-xm + y$ to $-xm + y + 1$. These limits are parallel lines in the m-b

plane with integer slope $(-x)$ and integer intercept. (The x-y and m-b spaces are dual to each other, with points in one corresponding to lines in the other [19].) The network of all such lines for $x = 0,1,...,n$ partitions the m-b plane into *facets* (Figure 10). For every x, every segment (7) with m and b in the interior of one facet will pass between the same two lattice points (x,y) and $(x,y+1)$. Thus, from Definition 5, every such segment has the same chain code. The open facets together with appropriate boundary points are the equivalence classes we seek.

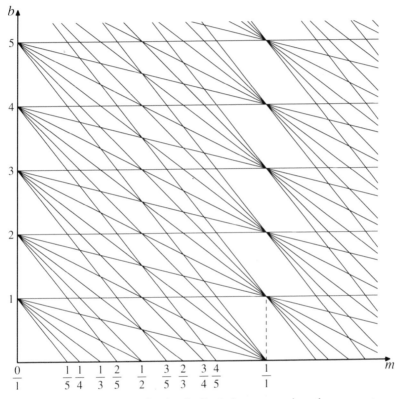

Figure 10. A Farey fan of order 5. Each facet comprises the parameters (m,b) of a family of lines $y = mx+b$ that have the same digital approximation over the interval $0 \le x \le 5$. The dotted line closes the basic rectangle $[0,1] \times [0,1]$.

The facet boundaries, lines of integer slope, fan out from integer points on the y axis. The m-intercepts between 0 and 1 form a Farey series, a well-studied object of number theory.* It may be verified that the ordinates of successive intersections

---

* A Farey series of order n is the sequence of all proper fractions with denominators less than or equal to n arranged in ascending order. The labels on the m-axis of Figure 10 form a Farey series of order 5.

along each ray also form Farey series of different orders. And the three distinct m-coordinates of each facet are successive terms of some Farey series. Consequently the figure has been dubbed a *Farey fan* [26].

Theorem 5. No facet in a Farey fan has more than four sides [9].

A simple explanation for Theorem 5 is that every segment in an equivalence class is a convex combination of segments in at most four extreme positions, which correspond to the vertices of the facet. Figure 11 exemplifies the four positions.

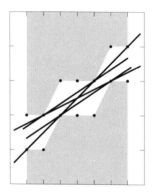

Figure 11. The top and bottom edges of the unit-height channel are 6-unit digital segments of slope 1/3. All lines that pass through the channel without touching the top have the same chain code. Vertical scale exaggerated.

The Farey fan of order n can be built by adding lines of slope $-n$ to a fan of order $n-1$. The values of m where the n new lines cut old lines within the basic rectangle, $0 \le m \le 1$, $0 \le b \le 1$, are precisely a Farey series of order n with the first term, 0/1, deleted. This observation leads to

Theorem 6. The number of distinct chain codes for n-unit digital segments is

$$1 + \sum_{k=1}^{n} (n+1-k)\phi(n).$$

As usual, $\phi(n)$ is Euler's totient, the number of positive integers less than n that are relatively prime to n. The number of terms in a Farey series of order n is

$$1 + \sum_{k=1}^{n} \phi(n),$$

from which the theorem follows directly [8]. The first ten values of the formula in the theorem are 2, 4, 8, 14, 24, 36, 54, 76, 104, 136. Asymptotically, the number of distinct n-unit digital segments is n^3/π^2 [2].

The size of a facet in the Farey fan shows the uncertainty inherent in a digital segment. For example, to recover the parameters of a line dropped on an integer lattice, one can read off a bit of chain code, from there determine the facet it lies in, and thence estimate its parameters [8]. The uncertainty is variable; it is greatest for the big facets at $m = 0$ and $m = 1$. This reflects the fact that an arrow shot along an alignment of trees in a square orchard is likely to go farther when the alignment is in a principal direction; then the arrow's path (a segment) need not be very precisely positioned to fit between the trees.

On average, an n-unit digital segment can be used to estimate the position of a selected point on a straight line to about $1/n$ unit [2]. Such estimates have been used to locate points in digital satellite images to subpixel accuracy [1]. The least uncertainty in estimating the parameters of a digitized line occurs for lines with slope τ^{-1}, the reciprocal of the golden ratio. Because multiples of τ^{-1} are as uniformly distributed modulo 1 as possible [21], the edges of the channel through which the line threads as in Figure 11 cannot be far apart.

For more about digital lines and further pointers to the literature, see Dorst [8] and Bruckstein [6], both of whose work appears in a recent volume of the AMS Contemporary Mathematics series [28]. For a generalization to 3-space, see Forchhammer [13].

5. Conclusion

Simple problems in computer graphics have been tamed by number-theoretic analysis. It is a routine and inexpensive matter to represent straight lines, standard circles, and standard ellipses to the ultimate precision of digital media. Artifacts of the representations (jaggies) are understood in detail. At least where straight lines are involved, it is possible to recover line data from digital images to subpixel accuracy. In another problem, that of drawing circles, the relationship among various approximation criteria has been characterized in some detail.

In computer graphics, more than in most numerical computing, one is vividly confronted by the discrete nature of the pursuit. There is an ultimate, finite level of precision. At that level, numerical analysis merges with number theory, and the imperfection of rounding fades into the exactness of integer computation. Such ultimate precision is occasionally approached in numerical analysis, for example in the exact treatment of Newton's method for the square root [18], in the ellipsoid method for linear programming [30], or in the analysis of floating-point base conversion [7]. It is interesting to speculate about the extent to which numerical computing may one day become explicable as a branch of number theory.

I am grateful to J. A. Reeds for helpful criticism and to J. C. Lagarias for many pointers to the literature.

REFERENCES

1. C. A. Berenstein, L. N. Kanal, D. Lavine, and E. C. Olson, *A geometric approach to subpixel registration accuracy,* Computer Vision, Graphics, and Image Processing **40** (1987), 334-360.

2. C. A. Berenstein and D. Lavine, *On the number of digital straight line segments,* IEEE Transactions on Pattern Analysis and Machine Intelligence **10** (1988), 880-887.

3. M. Boshernitzan and A. S. Fraenkel, *A linear algorithm for nonhomogeneous spectra of numbers,* Journal of Algorithms **5** (1984), 187-198.

4. J. Bresenham, *A linear algorithm for incremental digital display of circular arcs,* Communications of the ACM **20** (1977), 100-106.

5. R. Brons, *Linguistic methods for the description of a straight line on a grid,* Computer Graphics and Image Processing **3** (1974), 48-62.

6. A. M. Bruckstein, *Self-similarity properties of digitized straight lines,* department of Computer Science report #616, Technion – Israel Institute of Technology, March, 1990. Also in Melter, *Vision Geometry.*

7. W. D. Clinger, *How to read floating point numbers accurately,* ACM SIG-PLAN '90 Conference on Programming Language (White Plains, 1990) Design and Implementation, Association for Computing Machinery, 1990, pp. 92-101.

8. L. Dorst, *Discrete Straight Line Segments: Parameters, Primitives and Properties,* PhD Thesis, Technische Hogeschool Delft, 1986. Also in Melter, *Vision Geometry.*

9. L. Dorst and A. W. M. Smeulders, *Discrete representation of straight lines,* IEEE Transactions on Pattern Analysis and Machine Intelligence **6** (1984), 450-462.

10. H. Eves, *Elementary Matrix Theory,* Dover, 1966.

11. E. L. Fiume, *The Mathematical Structure of Raster Grpahics,* Academic Press, 1989.

12. J. D. Foley, A. Van Dam, S. K. Feiner, and J. F. Hughes, *Computer Graphics Principles and Practice,* Addison-Wesley, 1990.

13. S. Forchhammer, *Digital plane and grid point segments,* Computer Vision, Graphics and Image Processing **47** (1989), 373-384.

14. H. Freeman, *Computer processing of line-drawing images,* Computing Surveys **6** (1974), 57-97.

15. R. L. Graham, S. Lin, and C-S. Lin, *Spectra of Numbers,* Mathematics Magazine **51** (1978), 174-176.

16. G. H. Hardy and E. M. Wright, *An Introduction to the Theory of Numbers,* Oxford University Press, 1968.

17. B. K. P. Horn, *Circle generators for display devices,* Computer Graphics and Image Processing **5** (1976), 280-288.

18. A. S. Householder, *Principles of Numerical Analysis,* McGraw-Hill, 1953, pp. 10-13.

19. F. Klein, *Elementary Mathematics from an Advanced Standpoint,* Dover, 1945.

20. D. E. Knuth, *The Art of Computer Programming,* Vol. 2 Seminumerical Algorithms, Addison-Wesley, 1971, p. 320.

21. _____, *The Art of Computer Programming,* Vol. 3 Searching and Sorting, Addison-Wesley, 1973, p. 511.

22. J. Koplowitz, M. Lindenbaum, and A. Bruckstein, *The number of digital straight lines on an N×N grid*, IEEE Transactions on Information Theory **36** (1990), 192-197.

23. Z. Kulpa, *On the properties of discrete circles, rings, and disks*, Computer Graphics and Image Processing **10** (1979), 348-365.

24. Z. Kulpa and M. Doros, *Freeman digitization of integer circles minimizes the radial error*, Computer Graphics and Image Processing **17** (1981), 181-184.

25. M. D. McIlroy, *Best approximate circles on integer grids*, ACM Transactions on Graphics **2** (1983), 237-263.

26. _____, *A Note on Discrete Representation of Lines*, AT&T Technical Journal **64** (1985), 481-490.

27. _____, *Getting raster ellipses right*, ACM Transactions on Graphics (to appear).

28. R. A. Melter, A. Rosenfeld, and P. Bhattacharya (eds.), *Vision Geometry.* American Mathematical Society, Providence, 1991. Contemporary Mathematics Series **119**.

29. T. Nagell, *Introduction to Number Theory*, Chelsea, 1964.

30. C. H. Papadimitriou and K. Steiglitz, *Combinatorial Optimization: Algorithms and Complexity*, Prentice-Hall, 1982.

31. T. Pavlidis, *Algorithms for Graphics and Image Processing*, Computer Science Press, Rockville, MD, 1982.

32. C. Series, *The geometry of Markoff numbers*, The Mathematical Intelligencer **7**, 3 (1985), 20-29.

33. L. D. Wu, *On the chain code of a line*, IEEE Transactions on Pattern Analysis and Machine Intelligence **4** (1982), 347-353.

SOFTWARE AND SYSTEMS RESEARCH DEPARTMENT, AT&T BELL LABORATORIES, MURRAY HILL, NEW JERSEY, 07974

E-mail: doug@research.att.com

Index

Recent Titles in This Series

(*Continued from the front of this publication*)